Weakest Bound Electron Theory and Applications

Neng-Wu Zheng

Weakest Bound Electron Theory and Applications

Neng-Wu Zheng
Department of Chemistry
University of Science and Technology
of China
Hefei, China

ISBN 978-981-19-6656-9   ISBN 978-981-19-6657-6 (eBook)
https://doi.org/10.1007/978-981-19-6657-6

Jointly published with Shanghai Scientific and Technical Publishers
The print edition is not for sale in China (Mainland). Customers from China (Mainland) please order the print book from: Shanghai Scientific and Technical Publishers.

Translation from the Chinese language edition: "Zuiruoshouyueshudianzililunjiyingyong" by Neng-Wu Zheng, © The Author 2011. Published by University of Science and Technology of China Press. All Rights Reserved.
© Shanghai Scientific and Technical Publishers 2023
This work is subject to copyright. All rights are solely and exclusively licensed by the Publisher, whether the whole or part of the material is concerned, specifically the rights of reprinting, reuse of illustrations, recitation, broadcasting, reproduction on microfilms or in any other physical way, and transmission or information storage and retrieval, electronic adaptation, computer software, or by similar or dissimilar methodology now known or hereafter developed.
The use of general descriptive names, registered names, trademarks, service marks, etc. in this publication does not imply, even in the absence of a specific statement, that such names are exempt from the relevant protective laws and regulations and therefore free for general use.
The publishers, the authors, and the editors are safe to assume that the advice and information in this book are believed to be true and accurate at the date of publication. Neither the publishers nor the authors or the editors give a warranty, expressed or implied, with respect to the material contained herein or for any errors or omissions that may have been made. The publishers remain neutral with regard to jurisdictional claims in published maps and institutional affiliations.

This Springer imprint is published by the registered company Springer Nature Singapore Pte Ltd.
The registered company address is: 152 Beach Road, #21-01/04 Gateway East, Singapore 189721, Singapore

Preface

The appearance of things is complicated, but the essence is simple. It has been approved by many scientific theorems, rules, principles and theories. The author always adheres to this idea in order to reach the essence of things during the construction of Weakest Bound Electron Theory (WBE Theory) and the writing of this book.

The author introduced the idea of the weakest bound electron into theoretical chemistry, and constructed the WBE theory based on the wave-particle duality. It not only satisfies the requirements for the indistinguishability of identical particles and Pauli Exclusion Principle, but also underlines the properties of single particle and provides a theoretical basis for approximate separability between particles.

Hydrogen, one of the single-electron systems which are exactly soluble in quantum mechanics, has an electron which is the weakest bound electron in its system; therefore, it should be included in the theory.

It is very difficult to incorporate the wave-particle duality into the quantum theory of electron structures, and re-produce the properties and rules of atoms and molecules. This has been indicated by the establishment and development of all kinds of quantum theories and methods. Although currently the WBE theory and applications are just a framework, there is no doubt that it not only has to include, expand and improve the achievements from all available quantum theories and methods, but also make communications between different theories and methods based on wave-particle duality.

The author hopes that the WBE theory would bring new information and make contributions for the development and application of quantum theory.

Thanks to the students in my laboratory. They not only give me the joy of working together, but also make contributions to the development and application of the theory. Thank the older generation of scientists and friends for their help in my research. They are Mr. Ao-Qing Tang (唐敖庆, Academician of Chinese Academy of Sciences), Mr. Guang-Xian Xu (徐光宪, Academician of Chinese Academy of Sciences), Mr. Le-Min Li (黎乐民, Academician of Chinese Academy of Sciences), Prof. Ke-Min Yao (姚克敏, Zhejiang University), Prof. Xiang-Lin Zhang (张祥林, Central South University), Prof. Qian-Shu Li (李前树, Beijing University of Technology), Prof. Yao-Quan Chen (陈耀全, Chinese Academy of Sciences), Prof. Xiao-Yin Hong (洪啸吟, Tsinghua University), Prof. Han-Bao Feng (冯汉保, National Natural Science Foundation of China), Prof. Xiang-Lin Jin (金祥林, Peking University), Prof. Jia-Ju Zhou (周家驹, Chinese Academy of Sciences) and Rui-Shu Wang (王瑞书, Editor-in-chief of Jiangsu Phoenix Education Publishing House). Thank Ms. Zhe-Feng Gao (高哲峰) and other staffs from the University of Science and Technology of China Press for their effort to make this book publish. Finally, I am deeply grateful to my parents, my wife You-Xian Xu (徐幼仙) and children for their love, support and assistance.

Hefei, China   Neng-Wu Zheng

Acknowledgements With this monograph, I would like to express my memory for my parents forever.

Preface to the English Translation

The Chinese revised version of this monograph was published in 2011. The book systematically and exhaustively expounds a novel and original quantum theory—Weakest Bound Electron Theory (WBE Theory), which was first proposed and established by the author in the world. After the monograph was published, it won many honors and was deeply praised by the academic community. Ms. Lan Hua (花兰) devoted a lot of effort to translate the revised Chinese edition edited by Zhe-Feng Gao (高哲峰) and published by University of Science and Technology of China Press into English and delivered it to Shanghai Science and Technology Press. The publisher was so impressed with the original book and its social impact that it recommended it to Springer. The two publishers did a lot of reviewing and writing guidance to make this English version available. For this, the author would like to express heartfelt thanks to Ms. Lan Hua (花兰), Prof. Yi Cui (崔屹, Academician of the American Academy of Sciences), Ms. Dong-Xia Ma (马东霞, a scientist at the Max-Planck Institute), Prof. Guang-Can Guo (郭光灿, Academician of the Chinese Academy of Sciences) and Mr. Chen Zhang (张晨), Mr. Ying-Ming Ji (季英明) from Shanghai Scientific and Technical Publishers and Springer.

In the ten years since the Chinese version was published, the author has made many new thoughts and breakthrough research on WEB Theory and its application. Taking this opportunity, the author summarizes these new achievements into the following five points.

1. Basic ideas and prospects of WBE Theory

The proposal of quantum mechanics is one of the most significant scientific achievements, which profoundly affects the development of various branches of natural science and the progress of science and technology, and greatly expands people's cognition of the micro world. After quantum mechanics was proposed, many scientists developed a variety of approximate quantum theories and computational methods for dealing with multi-particle systems such as Hartree SCF method (Hartree SCF), Hartree-Fock SCF method (HF SCF), Molecular Orbital Theory (MO Theory) and Ab Initio method, Many-Body Perturbation Theory (MBPT), Frontier Orbital Theory, Quantum Monte Carlo method (QMC), Density Functional Theory (DFT),

Effective Core Potential (ECP), Quantum Mechanics/Molecular Mechanics method (QM/MM), Configuration Interaction (CI) and Coupled Cluster Theory (CCT). Some scientists engaged in research have won Nobel Prizes for their outstanding contributions.

The author's research is also aimed at establishing a new quantum theory. After years of research, the author finally put forward and established a novel and original innovative quantum theory for the first time in the world. The theory is named as Weakest Bound Electron Theory (abbreviated as WBE Theory). My students and I have applied the theory to (1) The study of atomic properties, including atomic energy level, oscillator strength, transition probability, radiation lifetime, etc. A large number of research results show that the WBE Theory has high accuracy, unity, simplicity and universality in these applications. (2) The study of energy levels and ionization energy. On the basis of defining the concept of iso-spectrum-level series, we propose the ionization energy difference law for ground and excited states of atoms. We also put forward the concept of spectrum-level-like series and a new formula for calculating atomic energy levels. (3) Molecular design research. The concept of nuclear potential of Weakest Bound Electron (WBE) is proposed and related to electronegativity and hard and soft acid base theory. Many coordination polymers with novel structures have been synthesized. Earlier, the design and synthesis of lanthanide coordination polymers were carried out internationally. (4) Taking the treatment of helium atom and helium-like ion series as an example, the accuracy and simplicity of WBE Theory in dealing with multi-electron atom system are demonstrated. (5) Taking molecular hydrogen ion treatment as an example, the feasibility of WBE Theory in dealing with molecular problems and the method that the delocalized molecular orbitals of WBE can be formed by a linear combination of corresponding atomic orbitals of WBE is demonstrated.

The novel ideas, clear physical nature and images, strict mathematical solutions and analytical expressions, as well as compatibility with existing concepts, principles, laws, theories and methods of quantum mechanics and quantum chemistry, all these advantages of WBE Theory will surely give WBE Theory a strong vitality.

The basic ideas of WBE Theory are briefly summarized as follows:

(1) The concept of WBE is introduced into quantum theory for the first time. WBE is the most easily excited and ionized electron in the current system, and also the most active electron in physical, chemical and biological processes. We introduce the concept of WBE into quantum theory for the first time and establish a new quantum theory.

(2) The N-electron atom problem can be simplified into N one-electron problems of WBE.

From the viewpoint of step-by-step ionization, the N electrons in the multi-electron system can be renamed as N WBEs. The electron's Hamilton operator of the system can be written as the sum of N one-electron Hamilton operators of WBE. The total electron energy of the system is equal to the sum of the N single-electron energies of WBE. It's also equal to the negative value of the sum of ionization energies step by step. Thus, the N electron atom problem can be reduced to N one-electron problems of WBE. In Hartree SCF, the electron's Hamiltonian of the system is actually changed due to repeated calculation of electron's repulsion energy. Therefore, the orbital energy is not single-electron energy, and the sum of orbital energy is not equal to the total electronic energy of the system.

(3) For multi-electron atoms, the single-electron Schrodinger equation of WBE can be solved strictly in the form of analytical potential. As we know, hydrogen atom and hydrogen-like ions are single-electron, single-center systems. The solution of hydrogen and hydrogen-like by quantum mechanics results in the radial wave function and energy expression marked by integer quantum number n, l and nuclear charge number Z. Imagine, if we introduce electrons into hydrogen and hydrogen-like, and form a new potential field with the nucleus, can n, l and Z in the radial wave function and energy expression of the original electron be integers again? In the multi-electron atom, each WBE moves in the potential field of the Non-Weakest Bound Electron (NWBE) of the current system and the nucleus. This is exactly the same as the above scenario. Based on this consideration, after careful study, we propose an expression of analytic potential. In this form, the single electron Schrodinger equation of WBE can be solved as strictly as the hydrogen atom system, and the analytical expressions of the energy and wave function of WBE corresponding to the hydrogen atom but n, l Z are non-integer (the radial wave function is the generalized Laguerre polynomial) can be obtained.

(4) Multi-center molecule problems can be simplified to a linear combination of single-center atom problems. With the introduction of a nucleus in a system of single-center atoms or ions, the potential field felt by electrons must change. This change is physically identical to the change in the potential field caused by the introduction of a new electron in single-center atoms or ions. And again, in a multi-nucleus system, there's a probability that the electron will be biased toward this nucleus or that nucleus, so you can express molecular orbitals in linear combinations. Thus, for molecular systems, WBE moves in a potential field consisting of all NWBEs and nuclei, and WBE molecular orbitals are essentially delocalized. It can be formed by a linear combination of suitable atomic orbitals of WBE.

(5) The process of removing and adding electrons provides two topologically equivalent modes for dealing with multi-electron problems.

For an N-electron system, the process of removing WBE one by one by successive ionization, and the process of adding electrons one by one to construct an N-electron system similar to the Aufbau process are the inverse process of each other. The two form a closed cycle. The idea of removing and adding electrons provides two topologically equivalent models for dealing with N-electron atoms and molecules. Both modes simply rename the electrons, so the Hamiltonian of the system remains the same. The two modes mean that electrons can be processed centrally or individually. It not only embodies the indistinguishability of identical particles and antisymmetric principle, but also finds a theoretical basis for the separable treatment of electrons which shows the personality of electron. In other words, many accepted physical ideas, concepts, principles and laws are observed in WBE Theory. For example, atomic orbital, molecular orbital, electron configuration, minimum energy principle, Pauli exclusion principle and conservation principle of orbital symmetry. It also provides insight into the behavior of individual electrons, such as the transition behavior of electron between energy levels. The calculations involved are also greatly simplified. In this respect, WBE Theory is quite different from the SCF method and model potential theory.

(6) Idea about simplifying calculations.

A large number of facts show that the interaction between distant atoms or atomic groups in a molecule is relatively small; in atoms or molecules, the inner electrons of atoms generally do not participate in many physical, chemical and biological processes (except for a few processes such as inner electron excitation). Some scientists have long been aware of these facts and have expressed them in their theories and methods, for example, Frontier Orbital Theory, Huckel Molecular Orbital Theory, Effective Core method, concept of molecular fragment and so on. On the basis of previous ideas, we have also incorporated these ideas, which are beneficial to greatly simplifying calculation and exploring the essence of the problem into WBE Theory. We suggest that the entire molecule can be approximately divided into one active fragment and one or/and several inert fragments, depending on the research intent, or/and the molecular structural properties. The effect of inert fragments on active fragment was treated in a constant quantitative or modeling manner to concentrate the active fragment. The active fragment can be roughly divided into outer active orbitals and electrons and inner inert orbitals and electrons. The inner inert orbitals and electrons can also be treated in a constant quantitative or modeling manner order to concentrate on the outer active orbitals and electrons. If necessary, the processing of the outer active orbitals and electrons can also be modeled and simulated by a computer. This idea about simplifying calculations, of course, is still just an imagine, and needs to be studied and confirmed.

Quantum mechanics and quantum chemistry have made remarkable achievements. Many ideas, concepts, principles, theories, laws and methods are emerging constantly. WBE Theory was put forward in this background. Its theory and application prospects have been preliminarily demonstrated and recognized by international peers. Just as Xun Lu (鲁迅) said that there was no road on the ground, when many people walk, it becomes a road. The author hopes and believes that in the field of quantum mechanics and quantum chemistry with abundant resources, WBE Theory will thrive and become a broad road with more people following it in the future.

2. Differences between Hamilton operators in Hartree SCF and WBE Theory

Since the birth of quantum mechanics, the quantum mechanical treatment of multi-electron system, namely the multi-body problem of quantum mechanics, has been paid much attention. So far, it is still impossible to find the exact solution of the

Schrodinger equation for a multi-particle system. Therefore, many scientists focus on finding an approximate solution of the Schrodinger equation for multi-particle system. Hartree SCF is one of the brightest lights in the research. On the basis of Hartree SCF, after improvement and development, the Hartree-Fock equation, Hartree-Fock-Roothaan equation, MO theory, ab initio method, MCSCF method and so on are produced. The Hartree-Fock equation has become one of the most important equations in quantum physics, condensed matter physics and quantum chemistry. It is also the basis of all quantum chemical calculation methods based on MO theory.

The basic idea of the Hartree SCF method is that each electron in the system moves independently in the average potential field of the nucleus and other electrons. Its single-electron Hamiltonian is

$$\widehat{H}_j = -\frac{1}{2}\nabla^2 - \frac{z}{r} + \sum_{i \neq j}^{N} \int \frac{1}{|r_i - r|} |\varphi_i(r_i)|^2 d\tau_i \tag{1}$$

The corresponding time-independent Schrodinger equation is

$$\widehat{H}_j \psi_j(r) = \varepsilon_j \psi_j(r) \qquad (j = 1, 2, \ldots, N) \tag{2}$$

The solution $\psi_j(r)$ is the single-electron state function of the electron j [1]. Then, a self-consistent solution is achieved by iterative method. Up to now, the improvement of Hartree SCF theory and method has only focused on wave function, and there are many good schemes and fruitful results have been achieved, such as HF SCF and so on. And my focus is on the single-electron Hamiltonian proposed by him. I have noticed that there are two crucial questions in the expression for the single-electron Hamiltonian of Hartree SCF. First, electron repulsive potential energy is repeatedly accounted for. As a result, the Hamiltonian of the system is changed. The orbital energy is not equal to the electron energy, and the sum of orbital energy is not equal to the total electron energy of the system. Second, because the potential function contains undetermined $\phi_i(r_i)$, the equation can not have an analytic solution. In this method, "self-consistency" can only be achieved by iterative approximation. These two deficiencies derived from the treatment of Hamiltonian bring inconvenience to the development of subsequent theoretical methods (including the impact on the universality, simplicity and accuracy of theoretical methods).

How does WBE Theory deal with Hamiltonian? The first way: according to the definition of step-by-step ionization, the N electrons in an N-electron atom are ionized one by one. The ionized electron is called the Weakest Bound Electron (WBE), and the unionized electron is called the non-Weakest Bound Electron (NWBE). Therefore, WBEi can be considered to move in a potential field consisting of (N-i) NWBEs and nucleus. Its one-electron Hamiltonian is

$$\widehat{H}_i = -\frac{1}{2}\nabla_i^2 - \frac{z}{r_i} + \sum_{i<j}^{N} \frac{1}{r_{ij}} \quad (3)$$

where i represents WBEi and j represents NWBEj. All repulsive potential energy terms between electron pairs associated with WBEi are assigned to WBEi. Obviously, the sum of the one-electron Hamiltonian of WBE is equal to the total electronic Hamiltonian of the system. The second way: it can also be expressed in a sentence. That is, the repulsive energy term of an interacting pair of electrons (identical particles) in the system should be amortized between the two electrons. Its one-electron Hamiltonian is

$$\widehat{H}_i = -\frac{1}{2}\nabla_i^2 - \frac{z}{r_i} + \frac{1}{2}\sum_{i\neq j}^{N} \frac{1}{r_{ij}} \quad (4)$$

Obviously, the sum of all one-electron Hamiltonian is equal to the total electronic Hamiltonian of the system.

Equations (3) and (4) are two ways of dividing multi-electron Hamiltonian into single-electron Hamiltonian proposed by the author according to the physical facts of step-by-step ionization and identical particles, and are also one of the core ideas of WBE Theory. Magnetic interaction terms between particles can be divided with the same idea.

For the "remove" mode mentioned above, the single-electron Hamiltonian takes the first way and for "Join" mode, it takes the second way. There are two things that are important in quantum mechanics and quantum chemistry. (1) The Hamiltonian of the system represents an observable physical quantity, namely the total energy of the system. (2) The standard state of quantum chemistry. Taken together, the total electronic energy of the system is equal to the negative of the sum of the successive ionization energies. In this way, the merits and demerits of a theory have the standard of experimental examination. Obviously, the above two ways of combining operators only rename the electrons from different angles. The sum of all single-electron Hamiltonian is equal to the total electron Hamiltonian of the system, and does not change the system's Hamiltonian. So the sum of the energies of the single electrons is equal to the total electron energy of the system and is equal to the negative of the sum of the successive ionization energies of the system. This solves the problem of overestimating electron repulsion energy in Hartree SCF.

In both ways, each electron moves in a potential field composed of a nucleus and other electrons, and the physical nature of a potential field is the same. Therefore, we propose an approximate unified potential function to express their potential fields:

$$V(r) = -\frac{z'}{r} + \frac{d(d+1) + 2dl}{2r^2} \quad (5)$$

In this way, the one-electron Schrodinger equation can be solved. The radial wave function of the solution is the generalized Laguerre polynomial (not the usual associated Laguerre polynomial).

In conclusion, the WBE Theory avoids the two major shortcomings of Hartree SCF's repeated calculation of the electron pair repulsion energy and the absence of analytical solutions to the single-electron Schrodinger equation.

3. Origin of potential function in WBE Theory

Everyone wondered how the multi-electron problem could be reduced to the single-electron problem, and how the single-electron problem could be solved analytically by solving the single-electron Schrodinger equation like the hydrogen atom problem. I've been working on this for years. The ionization of electrons in multi-electron systems is an important physical property of atoms and molecules. Ionization has a strict physical definition, and ionization energy is a physical quantity that can be measured experimentally. The Hamiltonian in quantum mechanics is directly related to this physical quantity.

Therefore, the concept of WBE and the change rule of the successive ionization energy of elements first came into my research sight. Firstly, I define the concept of iso-spectrum- level series and put forward the ionization energy difference law for ground and excited states of atoms: in an iso-spectrum-level series, the first order difference of ionization energy has a good linear relationship to the nuclear charge number. The second order difference is close to a fixed value. Only one WBE ionizes in each stage of ionization, so WBE can be considered to be moving in a potential field composed of all NWBEs and nucleus. No matter how many number of NWBEs is in this potential field, whether they are distributed in open shell or closed shell, the ionization energy difference law is obeyed. Therefore, I believe that the potential field felt by WBE can have a unified expression. An infinite series expression of the potential function of the monovalent electron potential field of alkali metals has been proposed. Can we introduce this expression directly? The answer is no. Because in infinite series form, the single-electron Schrodinger equation can't be solved. Moreover, the researchers only focused on the shielding and polarization effects of static electricity and did not consider the penetration effect of electron probability distribution. After studying the quantum mechanics of hydrogen atom and hydrogen spectrum, the infinite series expression of the monovalent electron potential function of alkali metals, the spectrum of alkali metals and the ionization energy difference law of elements, we finally proposed the analytic form of the potential function of WBE under the non-relativistic and central field approximation:

$$V(r_\mu) = -\frac{z'_\mu}{r_\mu} + \frac{d_\mu(d_\mu + 1) + 2d_\mu l_u}{2r_\mu^2} \tag{6}$$

Omit the subscript

$$V(r) = -\frac{z'}{r} + \frac{d(d+1) + 2dl}{2r^2} \tag{7}$$

In this form, the single-electron Schrodinger equation of WBE can be solved as strictly as the hydrogen atom equation. The angular part of the solution is the same as the hydrogen atom, i.e. spherical harmonics, and the radial equation can be solved by two methods, the generalized Laguerre function method and the gamma function method. If the generalized Laguerre function method is used to solve the problem, then the radial wave function obtained is a generalized Laguerre polynomial (instead of the usual associated Laguerre polynomial) [2]. Moreover, WBE theory can restore all the results of quantum mechanics treatment of a hydrogen atom. Therefore, WBE theory is a unified quantum theory of one-electron and multi-electron systems.

4. The WBE Theory applies to molecular systems

The basic idea is as follows:

(1) According to the definition of step-by-step ionization of molecule, only one WBE is removed from each step of ionization, while the remaining electrons (called NWBEs) are not removed. Therefore, every electron in the system qualifies as a WBE. According to the ionization view, WBE moves in a potential field composed of NWBEs and nucleus in the current system. Thus, the molecular orbitals of WBE are essentially delocalized.

(2) The multi-centric molecule problem can be simplified to the linear combination of the single-center atom problem, that is, the molecular orbitals can be linearly combined with the appropriate atomic orbitals of WBEs. The reason is that as a perturbation source, NWBE in the atom must affect the potential field generated by the nucleus felt by WBE, similarly for a molecule the addition of another nucleus besides the designated nucleus must also affect the potential field generated by the designated nucleus felt by WBE. The physical nature of the two is the same. Furthermore, in the current system, WBE can be close to the designated nucleus or any other nucleus, and WBE's motion is probability distribution.

Use WBE Theory to deal with hydrogen molecule ion (ground state) as a specific example: the simplest case can be selected

$$\psi = c_1 \varphi_a(1s) + c_2 \varphi_b(1s) \tag{8}$$

As a trial variational function, $\varphi_a(1s)$ and $\varphi_b(1s)$ are the WBE's 1s orbitals of the atom a and the atom b (not the 1s orbital of hydrogen atom) [2]. If readers are interested, they can refer to literature [3, 4] to continue the calculation by themselves.

5. Several research directions of WBE Theory

As WBE Theory is recognized and studied by more people, especially the involvement of interdisciplinary researchers in physics (including astrophysics, atomic physics, physical optics and quantum mechanics), chemistry (including inorganic chemistry, inorganic materials chemistry and quantum chemistry), biomedical (including medical technology, medical diagnosis and treatment), etc., the author predicts that WBE Theory may achieve results in the following four directions.

(1) The comparative study: Based on the advantages that the expression of the one-electron Hamiltonian in WBE Theory, the analytic form of the strictly solved radial wave function as the generalized Laguerre polynomial and the delocalized molecular orbitals of WBE can be linearly combined by the atomic orbitals of WBE, through the contrastive study with the existing quantum chemistry theories and calculation methods, it is possible to achieve results that change the current pattern of quantum chemistry and quantum mechanics.

(2) Research on new energy and new technologies: Chemical reactions, biological reactions and utilization of mineral energy only involve the change of the valence electron layer and the d and f electrons in the inner layer, while nuclear energy (nuclear fission and fusion) only involves the nucleus. However, the enthusiasm of low-level-energy electrons in large inner layers is not mobilized, little research. Based on the ionization energy difference law of elements proposed by the author as well as the calculation results [5] and WBE Theory, the research in this area may contribute to the development of new energy, new medical technology and diagnosis as well as treatment of some diseases (X-ray is a known example).

(3) Laser and physical optics: Based on the new energy level calculation formula proposed by us and the simple and accurate calculation method of WBE Theory in energy level, oscillator strength, transition probability, radiative lifetime, the research on energy level and energy level transition will make new contributions to laser and light source (especially new energy saving light source).

(4) Chemical symthesis and new materials: The combination of nuclear potential in WBE Theory with electronegativity hard and soft acid-base theory will play a great role in guiding material chemistry and inorganic chemical synthesis.

(5) The study of molecular systems and the study of catalytic transition states.

Hefei, China
January 2022

Neng-Wu Zheng (郑能武)

References

1. Xu GX (徐光宪), Li LM (黎乐民), Wang DM (王德民) (eds) (2009) A basic theory and Ab Initio calculation of quantum chemistry. Chinese volume. Science Press, Beijing, 4th printing, pp 153–155
2. Zheng NW (郑能武) (2011) Weakest bound electron theory and its application, (Revised edn.) [M], University of Science and Technology of China Press, Hefei
3. Yang ZD (杨照地), Sun M (孙苗), Yuan DD (苑丹丹), Zhang GL (张桂玲) (2012) Fundamentals of quantum chemistry. Chemical Industry Press, Beijing, p 77
4. Griffths DJ, translated by Jia Y (贾瑜), Hu X (胡行), Li YX (李玉晓) (2009) Introduction to quantum mechanics, translated edn. China Machine Press, Beijing, p 199
5. Zheng NW (郑能武) (1988) A new outline of atomic theory [M]. Jiangsu education publishing house, Nanjing

About This Book

When you read this book, you will feel a world permeated with the fragrance of innovation and enjoy the pleasure of a journey leading to the law and essence behind complex natural phenomena. Difference law for describing ionization energy, new formula for calculating energy levels, new perspective for reorganizing operator, unique thinking for constructing theory … all of these form a brand-new picture. A new quantum theory of original innovation is presented to the readers in this monograph, namely "Weakest Bound Electron Theory" proposed and constructed by Prof. Neng-Wu Zheng. Readers can systematically understand the content of the new theory and its broad application prospects in physics, chemistry, materials science and other fields, as well as learn scientific research methods and innovative spirit.

This monograph is suitable for those at the above undergraduate level (from undergraduates to scientists) and with an interest in scientific innovation.

Contents

1 **The Basics of Quantum Mechanics for the Weakest Bound Electron (WBE) Theory** 1
  1.1 The Wave-Particle Duality 1
  1.2 The Uncertainty Principle 1
  1.3 The Schrodinger Equation 3
  1.4 Electron Spin and Spin Orbital [3, 6–8] 6
  1.5 The Indistinguishability of Micro Identical Particles 9
  1.6 Pauli Exclusion Principle and Periodic Table 10
  1.7 One of the Approximation Methods in Quantum Mechanics—The Variation Method 14
  References 18

2 **The Weakest Bound Electron Theory (1)** 21
  2.1 The Concept of the Weakest Bound Electron 21
  2.2 Ionization Process and Aufbau-Like Process is Reversible 23
  2.3 The One-Electron Hamiltonian for the Weakest Bound Electron 26
    2.3.1 The Non-Relativistic One-Electron Hamiltonian for the Weakest Bound Electron 26
    2.3.2 The Treatment of Magnetic Interaction Between Electrons 30
    2.3.3 Relativistic Hamiltonian 31
  2.4 The One-Electron Schrodinger Equation of the Weakest Bound Electron 33
  2.5 The Key Points of the WBE Theory 35
  References 35

3 **The Weakest Bound Electron Theory (2)** 37
  3.1 Potential Function 37
  3.2 The Solution of the Radial Equation 39
    3.2.1 Spherical Harmonic 39
    3.2.2 Generalized Laguerre Functions 42

xix

	3.2.3 Restore the Form of Hydrogen and Hydrogen-Like Atoms	47
	3.2.4 The Definition and Properties of Generalized Laguerre Functions	48
	3.2.5 The Proof of the Satisfaction of Hellmann–Feynman Theorem	54
3.3	Matrix Element and Mean Value of Radial Operator r^k	56
3.4	The Exact Solutions of Scattering States in WBEPM Theory	58
3.5	The Formula for the Calculation of Fine Structure	60
3.6	Calculation of Spin–Orbit Coupling Coefficient	61
3.7	Relation Between the WBEPM Theory and Slater-Type Orbitals	62
References		66

4 The Application of the WBE Theory ... 69

4.1	Ionization Energy [1–10]	69
	4.1.1 Introduction	69
	4.1.2 Iso-spectrum-level Series and the Differential Law of Ionization Energy in the Series	76
	4.1.3 Calculation of Ionization Energy	86
	4.1.4 The Successive Ionization Energies of the $4f^n$ Electrons for the Lanthanides [10]	91
4.2	Energy Level [39–50]	96
	4.2.1 Introduction	96
	4.2.2 Formulae for Calculating Energy Levels	99
	4.2.3 Methods for Parameter Characterization	101
	4.2.4 Examples	107
4.3	Calculation of Oscillator Strength, Transition Probability and Radiative Lifetime [88–104]	129
	4.3.1 Introduction	129
	4.3.2 Theory and Method for Calculation	131
	4.3.3 Examples	135
4.4	Calculation of Total Electron Energy [1, 159, 160]	155
	4.4.1 Calculation of Total Electron Energy of the System Using Ionization Energy	157
	4.4.2 Variational Treatment on the Energy of the He-Sequence Ground State with the WBEP Theory	158
	4.4.3 Perturbation Treatment on the Energy of the He-Sequence Ground State with the WBEPM Theory [160]	176
4.5	Electronegativity, Hard and Soft Acids and Bases, and the Molecular Design of Coordination Polymers	179
	4.5.1 The Electronegativity Concept and Scale	179

4.5.2	The Nuclear Potential Scale of the Weakest Bound Electron [185, 200]	180
4.5.3	The Hard-Soft-Acid-Base Concept and Scale	185
4.5.4	Molecular Design of Coordination Polymers	188
References		196

Representation Publications 207

Postscript 211

Index 213

About the Author

Dr. Neng-Wu Zheng is a distinguished professor of theoretical inorganic chemistry and coordination polymers, notable for pioneering the "Weakest Bound Electron Theory". Major contributions: (1) Proposed a new quantum Theory (WBE Theory) internationally; (2) The difference law of ionization energy of elements is proposed internationally.

Brief Introduction of the Author

Neng-Wu Zheng, came from Wuyishan city (originally Chong'an County), Fujian Province. After graduation from Chong'an first middle school, he was admitted to the Department of Chemistry, Peiking University, and then graduated (six years are required at school). Now he is the professor and Ph.D. supervisor at the University of Science and Technology of China (USTC), and was honored with Special Government Allowance of the State Council of China. He has been mainly working on the field of theoretical inorganic chemistry and coordination polymers. He first proposed the Weakest Bound Electron Theory in the world. He published three monographs and dozens of research articles related to the theory, which attracted considerable attention from national and international peers, in famous international journals. He gave lectures on "Advanced Inorganic Chemistry" and "Inorganic Chemistry". By collaborating with others, he published five textbooks and teaching reference books. He also published four popular science books. He visited Purdue University as a visiting scholar. He served as the dean of the Department of Applied Chemistry, university academic committee member and university academic degree member at USTC. He was member of International Advisory Board of Malaysia J. Chem. Two of the Ph.D. and master's students he mentored won the President's Award of CAS and his students are all around the world.

Chapter 1
The Basics of Quantum Mechanics for the Weakest Bound Electron (WBE) Theory

1.1 The Wave-Particle Duality

The phenomenon of optical interference, diffraction, polarization, etc., shows that light is composed of waves; while the phenomenon of blackbody radiation, the photoelectric effect, etc., demonstrates that light consists of particles. Thus, light exhibits the properties of both waves and particles.

Inspired by light's wave-particle duality, in 1923, the French physicist L.V. de Broglie proposed a hypothesis, claiming that all matter (particles with non-zero rest mass $m \neq 0$, such as electrons, atoms, molecules, etc.) has a wave-like nature, and gave the famous formula as follows

$$\lambda = \frac{h}{p} = \frac{h}{mv} \tag{1.1.1}$$

This formula is known as De Broglie's formula, which relates the momentum p with a particle nature to the wavelength λ with a wave nature. The wave related to material particles, is so called De Broglie's wave or matter wave.

De Broglie's hypothesis was confirmed later by a series of experiments including electron beam, helium atom beam, hydrogen molecular beam, etc. One of the most famous experiments is the observation of electron diffraction in an experiment performed by C. J. Davison and L. H. Germer.

1.2 The Uncertainty Principle

In the macro-world, the position and momentum of one object can be accurately measured simultaneously. Objects move along the determined paths (or trajectories) in the space, and such paths or trajectories obey the Newton law. Objects can be discriminated by tracking the exact path or trajectory that each object takes. However, in the micro-world, owning to wave-particle duality, there is a fundamental limit to the precision with which the position and momentum of a micro-particle can be measured simultaneously. In 1927, W. Heisenberg stated that it is impossible to precisely determine the position and momentum of a micro-particle at the same time. Its uncertainty satisfies the following formula.

$$\Delta p \Delta q \gtrsim \frac{\hbar}{2} \tag{1.2.1}$$

This is called the uncertainty relation between position and momentum of a micro-particle. In the above formula, $\hbar = h/2\pi$, where h is Planck constant; Δq represents a measure of uncertainty for a micro-particle's position; Δp represents a measure of uncertainty of its momentum simultaneously.

Due to the uncertainty relation between position and momentum of a micro-particle, it is impossible for the micro-particle to have an accurate path or trajectory like a macro object. Thus, the interacting particles of the same kind are indistinguishable during movement.

Similar to the uncertainty relation between position and momentum, it also exists between energy and time, which is

$$\Delta t \Delta E \gtrsim \frac{\hbar}{2} \tag{1.2.2}$$

This formula indicates that for a system to clearly change its state the uncertainty of needed time is Δt and the uncertainty of the energy is ΔE.

One example of indeterminacy between energy and time can be seen in atomic energy level. Since atoms at excited state can spontaneously transit to lower energy state, they are unstable. If we use Δt to represent the average lifetime for atoms at excited state, based on the uncertainty relation between energy and time, the energy level with average lifetime Δt should have a natural width ΔE. Therefore, a real atomic energy level cannot be represented by a single value. The smaller the width ΔE, the longer the average lifetime, and the more stable the energy level, i.e., the harder the spontaneous transition, and vice versa. Through experiments, we can obtain the width of energy level by measuring the energy of photons from spontaneous emission, and then infer the average lifetime of the energy level. The stableness of atomic energy level is strongly coupled with the phenomenon of spontaneous transition and laser formation [1].

The uncertainty relation of all kinds that above mentioned is a universal law in the micro world, therefore, it is named the uncertainty principle [1].

The uncertainty principle comes from wave-particle duality. In quantum mechanics, due to the uncertainty principle, the interacting particles of the same kind are indistinguishable during movement.

1.3 The Schrodinger Equation

In 1926, E. Schrodinger proposed the famous wave function (also known as Schrodinger equation), which reveals the fundamental law of the movement of micro-particles and is the basic function of quantum mechanics. In the system that Schrodinger equation describes, probability (or the number of particles) is conserved and the particle velocity is far smaller than that of light. The conservation of probability (or number of particles) means nonexistence of the creation and destruction of particles. The creation and destruction of particles exists in the high energy field of atomic nuclear decay and nuclear reaction, while it doesn't exist for most questions about atoms and molecules. Because the velocity of particles in the questions about atoms and molecules is far smaller than the velocity of light, the non-relativistic relation between energy (E) and momentum (p) is used in the Schrodinger equation. For free particles, the relation is

$$E = \frac{p^2}{2m}$$

and for particles in the potential field V, the relation is

$$E = \frac{p^2}{2m} + v$$

Therefore, the Schrodinger equation is non-relativistic.

The Schrodinger equation includes time-dependent form and time-independent form.

The general expression of the time-dependent Schrodinger equation is:

$$i\hbar \frac{\partial}{\partial t} \Psi(r, t) = \left[-\frac{\hbar^2}{2m} \nabla^2 + V(r, t) \right] \Psi(r, t) \qquad (1.3.1)$$

where $\Psi(r, t)$ is related not only to position but also to time.

In the above time-dependent Schrodinger equation, when potential $V = V(r)$, i.e., the potential energy function is only related to position but not to time, the energy of the system has a definite value. The states with definite energy are called stationary states, and generally the stationary-state Schrodinger equation is the equation describing this situation. The equation has a general form as follows:

$$i\hbar \frac{\partial}{\partial t} \Psi(r, t) = \left[-\frac{\hbar^2}{2m} \nabla^2 + V(r) \right] \Psi(r, t) \qquad (1.3.2)$$

$\Psi(r, t)$ in (1.3.2) is the stationary-state wave function. The system in stationary states has a series of important properties including that the spatial probability density of particles, the average value of mechanical quantities, etc. don't change with time.

The stationary Schrodinger equation, i.e. Eq. (1.3.2) can be solved by separation of variables, i.e. letting $\Psi(r, t) = \psi(r) f(t)$. Its specific solution is

$$\Psi(r, t) = \psi(r) f(t) = \psi(r) \exp\left(-\frac{iEt}{\hbar}\right) \quad (1.3.3)$$

where $\psi(r)$ is the solution of the following equation

$$\left[-\frac{\hbar^2}{2m}\nabla^2 + V(r)\right]\psi(r) = E\psi(r) \quad (1.3.4)$$

$\psi(r)$ is unrelated to time, thus Eq. (1.3.4) is the time-independent Schrodinger equation.

Since Eq. (3.1.4) describes a system where $V = V(r)$ as well, it has the same characteristics of the system of $V = V(r)$. So Eq. (1.3.4) is often called the stationary Schrodinger equation, and $\psi(r)$ is also called the stationary-state wave function. [2–5]

In Eq. (1.3.4), we let

$$\widehat{H} = -\frac{\hbar^2}{2m}\nabla^2 + V(r) \quad (1.3.5)$$

then we get the operator for time-independent Schrodinger equation

$$\widehat{H}\psi(r) = E\psi(r) \quad (1.3.6)$$

where H is the Hamilton operator, the energy operator of the system. E is the energy eigenvalue of the system, and the corresponding wave function $\psi(r)$ is the energy eigenfunction. Therefore, the time-independent Schrodinger equation (1.3.6) or (1.3.4) is actually the energy eigenvalue equation of the system.

Mathematically for any E value, Eq. (1.3.4) or (1.3.6) is soluble, but the obtained solutions can not necessarily satisfy the requirements in physics. The solutions (i.e. wave functions) which satisfy the physical requirements must be of single-valuedness, boundedness and continuity in the variation range of all its variables.

For most physical and chemical problems related to atoms and molecules, the potential energy V is only coupled to position, and the potential and wave function do not change with time. All of them can be treated using Eq. (1.3.4). Therefore, the time-independent Schrodinger equation is very important.

How to write the Hamiltonian operator of (1.3.6) under certain representation and the specific expression of the corresponding Schrodinger equation? For atoms having N electrons and molecules having N electrons and X nuclei with fixed nuclear configuration, in the coordinate representation and Born-Oppenheimer approximation, the non-relativistic Hamiltonian operator of electron (atomic unit) is

1.3 The Schrodinger Equation

$$\hat{H}(1, 2, \cdots, N) = \sum_{\mu=1}^{N}\left(-\frac{1}{2}\nabla_{\mu}^{2}\right) - \sum_{A}^{X}\sum_{\mu}^{N} Z_A r_{A\mu}^{-1} + \sum_{v}^{N}\sum_{\mu<v}^{N}\frac{1}{r_{\mu v}} \quad (1.3.7)$$

where if $X = 1$, Eq. (1.3.7) is the Hamiltonian operator for atoms having N electrons; if $X > 1$, it represents the Hamiltonian operator for electrons in molecules having N electrons and X nuclei with fixed nuclear configuration.

Corresponding to this operator, the time-independent Schrodinger equation is

$$\hat{H}\psi(1, 2, \ldots, N) = E\psi(1, 2, \cdots, N) \quad (1.3.8)$$

$\psi(1, 2, \ldots, N)$, the solution of Eq. (1.3.8), describes the electron motion in atoms or in molecules with fixed nuclear configuration, and E is the total energy of electrons in the system.

Because Eq. (1.3.8) (or 1.3.4, or 1.3.6) is the energy eigenvalue equation of the system, Eq. (1.3.7) can also have the following form

$$\hat{H} = \hat{T} + \hat{V} \quad (1.3.9)$$

where $\hat{T}$ is the kinetic energy operator,

$$\hat{T} = -\frac{1}{2}\sum_{\mu=1}^{N}\nabla_{\mu}^{2} \quad (1.3.10)$$

and $\hat{V}$ is the potential energy operator,

$$\hat{V} = \sum_{A}^{X}\sum_{\mu}^{N}\left(-Z_A r_{A\mu}^{-1}\right) + \sum_{\mu}^{N}\sum_{\mu<v}^{N}\frac{1}{r_{\mu v}} \quad (1.3.11)$$

The first term on the right side of Eq. (1.3.11) represents the electron-nuclei attraction energy, and the second term represents the electron-electron repulsion energy.

For atomic system, E in Eq. (1.3.8) is the total energy of the system. While for molecules with fixed nuclear configuration, E represents the energy of electron under Born-Oppenheimer approximation. Except for E, the total energy of the system W should include the repulsion energy between nucleus and nucleus, which is

$$W = E + \sum_{A<B} Z_A Z_B R_{AB}^{-1} \quad (1.3.12)$$

W changes along with the changes in the nuclear configuration. It is one of the concerns in the study of potential energy surface.

1.4 Electron Spin and Spin Orbital [3, 6–8]

To explain the phenomena in atomic spectrum such as the doublet structures of the spectrum of alkali metals, Zemman effect, Stern-Gerlach experiment, etc. G. E. Uhlenbeck and S. Goudsmit postulated that the electron had spin in 1925.

Electron spin is different from its orbital motion.

Spin angular momentum of electron is $\hbar/2$, while orbital angular momentum is integer multiple of $\hbar$; electron spin magnetic moment divided by spin angular momentum is equal to e/mc, while electron orbital magnetic moment divided by orbital angular momentum is equal to $e/2mc$. So there is one fold difference between spin and orbital motion. In other words, Landè factor or g factor (gyromagnetic ratio) is $|g_s| = 2$ for electron spin, but $|g_l| = 1$ for electron orbit. These two properties of electron spin, which are different from orbital motion, result in the doublet structures of the spectrum of alkali metals and Zemman effect.

Spin angular momentum of a single electron can be represented by a vector operator s, and its component along x, y, z, axis is s_x, s_y, and s_z, respectively. The square of spin angular momentum commutes with the component s_x, s_y, and s_z

$$[s^2, s_x] = 0 \tag{1.4.1}$$

$$[s^2, s_y] = 0 \tag{1.4.2}$$

$$[s^2, s_z] = 0 \tag{1.4.3}$$

The eigenvalue of s^2 and s_z (component along z axis) are respectively

$$s(s+1)\hbar^2, \quad s = \frac{1}{2} \tag{1.4.4}$$

and

$$m_s \hbar, \quad m_s = \pm \frac{1}{2} \tag{1.4.5}$$

There are only two eigenstates, i.e. spin wave functions, for s^2 and s_z, represented by α and β.. Usually α and β are functions of magnetic spin quantum number $m_s m_s$

$$\alpha = \alpha(m_s), \quad m_s = +\frac{1}{2} \tag{1.4.6}$$

$$\beta = \beta(m_s), \quad m_s = -\frac{1}{2} \tag{1.4.7}$$

where α state is called spin-up state and β state is called spin-down state. The symbol ↑ and ↓ is used to represent α state and β state, respectively.

1.4 Electron Spin and Spin Orbital

In quantum mechanics, eigenfunctions need to be normalized. α and β are two eigenstates of Hermite operator s_z with different eigenvalues, and they are orthogonal. Thus, these two spin state are orthonormal

$$\sum_{m_s=-1/2}^{1/2} |\alpha(m_s)|^2 = 1 \tag{1.4.8}$$

$$\sum_{m_s=-1/2}^{1/2} |\beta(m_s)|^2 = 1 \tag{1.4.9}$$

$$\sum_{m_s=-1/2}^{1/2} \alpha^*(m_s)\beta(m_s) = 0 \tag{1.4.10}$$

To satisfy above orthonormal conditions, we must let

$$\alpha(1/2) = 1, \quad \alpha(-1/2) = 0 \tag{1.4.11}$$

$$\beta(1/2) = 0, \quad \beta(-1/2) = 1 \tag{1.4.12}$$

That electron has spin and corresponding magnetic moment is the intrinsic property of electron. The existence of this property indicates that electron has a new degree of freedom.

The wave function which describes the state of electron motion must include the intrinsic spin of electron which is unrelated to spatial motion. Therefore, one atomic orbital must be characterized by four different quantum numbers: n, l, m, m_s. n is the principle quantum number, n = 1, 2, 3, $\cdots$. l is azimuthal quantum number, l = 0, 1, 2, $\cdots$, n − 1. m is called magnetic quantum number, which represents the projection of the orbital angular momentum along the magnetic field, m = 0, ±1, ±2, $\cdots$, ±l. m_s is called spin quantum number, which represents the projection of the spin angular momentum along the magnetic field, $m_s = \pm 1/2$. To describe non-relativistic electrons under the central-field approximation, we select $\left(H, l^2, l_z, s_z\right)$ as a full set of conserved quantities. The common eigenstate is ψ_{nlmm_s}

$$\psi_{nlmm_s}(r, , m_s) = \psi_{nlm}(r)x(m_s) \tag{1.4.13}$$

Equation (1.4.13) indicates that the total wave function for a single electron is the product of the spatial orbital ψ_{nlm} and spin function $x(m_s)$. This wave function is called a spin orbital. For molecules, the definition of spin orbital is still the product of spatial orbital and spin function. But because their symmetry is different from that of atoms, the spatial orbital wave function is no longer labeled with nlm.

One spin orbital can only have one electron, but one spatial orbital can contain two electrons with opposite spin.

With the *LS* coupling, the energy states of a multi-electron atom can be classified by quantum number *S*, *L* and *J*.

S, *L* and *J* represent the total spin angular momentum, the total orbital angular momentum and the total angular momentum, respectively.

$$S = \sum s \quad (1.4.14)$$

$$L = \sum l \quad (1.4.15)$$

$$J = L + S \quad (1.4.16)$$

The above tree equations show that the total spin angular momentum *S* is the vector sum of spin angular momentum *s* of each electron; the total orbital angular momentum *L* is equal to the vector sum of angular momentum *l* of each electron; the total angular momentum *J* is equal to the vector sum of *L* and *S*. The value of *S*, *L* and *J* can be measured by *S*, *L* and *J*. If $L \geq S$, the total number of *J* is $2S+1$; if $L < S$, the number of *J* is only $2L+1$. So the energy state of atoms can be written as follows

$$^{2S+1}L_J \quad (1.4.17)$$

where $2S+1$ is the multiplicity of the term, ^{2S+1}L denotes the multi-electron states, and $^{2S+1}L_J$ represents the atomic energy level or a spectral branch. For *L*, there is a specific spectroscopic notation, i.e. $L = 0, 1, 2, \cdots$ denotes S, P, D, F, G, H, $\cdots$, respectively.

Except for the *LS* coupling, there is *jj* coupling, which merges *s* and *l* into *j* for each electron, then merges *j* into *J*; there is also *J'l* coupling, which combines all angular momentums except for spin s_i of the last electron into *K* and then combines *K* and s_i into the total angular momentum *J*.

Electron spin is not a classic effect, and its theoretical treatment falls into the category of relativistic quantum mechanics. Both electron orbital motion and spin motion have a magnetic moment, and the interaction between them is called spin-orbit coupling. The spin-orbit coupling energy is derived using relativistic quantum mechanics [5, 9], that is

$$E_{so} = \frac{1}{2\mu^2 c^2} \frac{1}{r} \frac{dV(r)}{dr} LS \quad (1.4.18)$$

Its corresponding operator is

$$H_{so} = \frac{1}{2\mu^2 c^2} \frac{1}{r} \frac{dV(r)}{dr} \hat{L}\hat{S} = \xi(r)\hat{L}\hat{S} \quad (1.4.19)$$

1.5 The Indistinguishability of Micro Identical Particles

In the microscopic world, the identical particles are the particles with inherent identical characters, for example electron. All electrons have exactly the same intrinsic properties such as mass (static mass is equal to 1 a.u., where 1 a.u. = 9.1091×10^{-28} g), electric charge (charge is equal to 1 a.u., where 1 a.u. = 4.80298×10^{-10} esu), spin (spin quantum number s = 1/2, a half-interger), etc. Thus, electrons are identical particles. The system with all identical particles is called the identical particle system. All electrons in an atom or a molecule constitute an identical particle system.

Due to the completely same intrinsic properties of identical particles, it is impossible to distinguish between them using differences in their intrinsic properties. As we mentioned in the previous section, identical particles are indistinguishable during motion because of the uncertainty principle. Therefore, based on the basic properties of the microscopic world-quantization and wave-particle duality of micro-particles, the interacting particles in an identical particle system is indistinguishable.

Since the interacting particles in an identical particle system can't differentiate between each other, the state of system doesn't change by exchanging all coordinates (including spatial coordinates and spin coordinates) of any two particles, or renaming or renumbering them; Any observable physical quantities of the system, especially Hamiltonian, are unchangeable [2, 9–11]. If we use q_i ($i = 1, 2, \cdots, i, \cdots, N$) to represent all the coordinates of particle i, the wave function $\Psi(q_1, \cdots, q_i, \cdots, q_j, \cdots, q_N)$ to represent the state of the identical particle system, and $\hat{P}_{ij}$ to represent the interchange of all coordinates of particle i and particle j, the state that $\hat{P}_{ij}\Psi(q_1, \cdots, q_i, \cdots, q_j, \cdots, q_N) = \Psi(q_1, \cdots, q_j, \cdots, q_i, \cdots, q_N)$ describes is the same state of the system since the two particles are identical.

The difference between $\hat{P}_{ij}\Psi(q_1, \cdots, q_i, \cdots, q_j, \cdots, q_N)$ and $\Psi(q_1, \cdots, q_i, \cdots, q_j, \cdots, q_N)$ is only a constant at most

$$\hat{P}_{ij}\Psi(q_1, \cdots, q_i, \cdots, q_j, \cdots, q_N) = c\Psi(q_1, \cdots, q_i, \cdots, q_j, \cdots, q_N) \quad (1.5.1)$$

Multiplying both sides of the above equation by the exchange operator $\hat{P}_{ij}$ from left, we get

$$\begin{aligned}\hat{P}_{ij}^2\Psi(q_1, \cdots, q_i, \cdots, q_j, \cdots, q_N) &= \hat{P}_{ij}c\Psi(q_1, \cdots, q_i, \cdots, q_j, \cdots, q_N) \\ &= c\hat{P}_{ij}\Psi(q_1, \cdots, q_i, \cdots, q_j, \cdots, q_N)\end{aligned} \quad (1.5.2)$$

Insert Eq. (1.5.1), then

$$\begin{aligned}\hat{P}_{ij}^2\Psi(q_1, \cdots, q_i, \cdots, q_j, \cdots, q_N) &= \hat{P}_{ij}c\Psi(q_1, \cdots, q_i, \cdots, q_j, \cdots, q_N) \\ &= c^2\Psi(q_1, \cdots, q_i, \cdots, q_j, \cdots, q_N) \quad (1.5.3)\end{aligned}$$

Using $\hat{P}_{ij}$ twice is equal to no net effect, so the square of $\hat{P}_{ij}$ is the unit operator, i.e. $\hat{P}_{ij}^2 = 1$, and then

$$c^2 = 1 \tag{1.5.4}$$

$$c = \pm 1 \tag{1.5.5}$$

This means that $\hat{P}_{ij}$ only has two eigenvalues. When $c = +1$, and this wave function is called symmetric wave function; when $c = -1$, and this wave function is called anti-symmetric wave function.

In other words, the indistinguishability of identical particles imposes strict limitations on the wave functions of the system. The exchange between any pair of particles in the identical particle system requires the wave functions of the system to be symmetric or anti-symmetric, and it doesn't change with time. This is called the principle of the indistinguishability of identical particles [4].

Experiments showed that the exchange symmetry of wave functions of identical particle system is associated with the spin of particles. For particles with integral spin ($s = 0, 1, 2, \cdots$), the system wave functions are always symmetric with respect to exchange of the particles, and these particles are called bosons; while for particles with half integral spin ($s = 1/2, 3/2, \cdots$), the system wave functions are always anti-symmetric with respect to exchange of the particles, and these particles are called fermions.

1.6 Pauli Exclusion Principle and Periodic Table

The electron spin ($s = 1/2$) is half integral, so the total electron wave function of the multi-electron system must be anti-symmetric with respect to exchange of all coordinates of any pair of electrons, i.e.

$$\begin{aligned}&\hat{P}_{ij}^2 \Psi(q_1, \cdots, q_i, \cdots, q_j, \cdots, q_N) \\ &= \Psi(q_1, \cdots, q_i, \cdots, q_j, \cdots, q_N) \\ &= -\Psi(q_1, \cdots, q_i, \cdots, q_j, \cdots, q_N)\end{aligned} \tag{1.6.1}$$

In quantum mechanics, this is known as the Pauli Exclusion Principle or anti-symmetric principle. It is one expression of the Pauli Exclusion Principle.

For an N-electron system, how to create the electron wave functions of the multi-electron system which satisfy the requirement of anti-symmetry?

As the Hamiltonian operator of an N-electron system includes a term for the electron-electron interactions, i.e. r_{ij}^{-1}, so far people haven't found a precise method by which the Hamiltonian of a multi-electron system can be written as a sum of one-electron Hamiltonians, as a result some approximations have been proposed.

1.6 Pauli Exclusion Principle and Periodic Table

One of the approximations is the independent particle approximation [9]. After neglecting repulsion between electrons, there is no interaction between electrons, and they can move independently. Then the Hamiltonian $\hat{H}$ of the system can be written as a sum of N one-electron Hamiltonians

$$\hat{H} = \sum_{i=1}^{N} H_i$$

According to the basic probability theory, the probability that multiple events occur independently at the same time is the product of the probability for a single event to occur. So the wave function can be expressed as the product of one-electron wave functions. However, the product of wave functions is neither symmetric nor anti-symmetric after exchanging all the coordinates of any pair of electrons, so anti-symmetrization must be applied. In quantum mechanics, the system wave function $\Psi^A(1, 2, \cdots, N)$ can be expressed as a Slater determinant which satisfies the requirement of antisymmetry.

In atoms or molecules, the repulsive interaction between electrons is very strong and has a very large contribution to the energy. Thus, neglecting the repulsive energy between electrons will lead to large errors. However, the independent particle approximation is very enlightening.

Another approximation is the orbital approximation [9, 10]. Since the multi-electron Hamiltonian includes the inter-electron repulsion r_{ij}^{-1}, it can't be simply written as the sum of one-electron Hamiltonians. Thus, it is approximated as a modified multi-electron Hamiltonian $\mathcal{F}(1, 2, \cdots, N)$, and $\mathcal{F}(1, 2, \cdots, N)$ can be written as the sum of "effective" one-electron Hamiltonians $F(i)$,

$$\mathcal{F}(1, 2, \cdots, N) = \sum_{i=1}^{N} F(i) = \sum_{i=1}^{N} \left[-\frac{1}{2} \nabla_i^2 + V(i) \right]$$

where $V(i)$ is the potential operator for electron i, and it represents the average potential field generated by the atomic nucleus and the rest $(N - 1)$ electrons.

When the operator $\mathcal{F}(1, 2, \cdots, N)$ is applied to Schrodinger equation, its solution can be written as the product of one-electron wave functions, i.e. the primitive Hartree product. The one-electron wave function ψ_i satisfies the one-electron Schrodinger equation

$$F_i(\mu)\psi_i(\mu) = \varepsilon_i \psi_i(\mu)$$

Then we apply antisymmetrization to the product wave function, which leads to the creation of the Slater determinant that satisfies antisymmetry requirement.

Thus, as long as the Hamiltonian of the system can be written as the sum of one-electron Hamiltonians, we can use the one-electron state function (or the one-electron spin orbital) as an element, and generate a fully antisymmetric electronic wave function for a multi-electron system through the form of the Slater determinant (single determinant or a linear combination).

For example, for an N-electron ($N = 2n$) system with every spatial orbital occupied by two electrons, the simplest expression for a single Slater determinant can be written

$$\Psi(1,2,\cdots,N) = \frac{1}{\sqrt{N!}} \begin{vmatrix} \psi_1(1)\alpha(1) & \psi_1(1)\beta(1) & \cdots & \psi_n(1)\beta(1) \\ \psi_1(2)\alpha(2) & \psi_1(2)\beta(2) & \cdots & \psi_n(2)\beta(2) \\ \vdots & \vdots & & \vdots \\ \psi_1(N)\alpha(N) & \psi_1(N)\beta(N) & \cdots & \psi_n(N)\beta(N) \end{vmatrix} \quad (1.6.2)$$

In the above determinant, all elements in each column have the same spin orbital and all elements in each row contain the same electron. $1/\sqrt{N!}$ is the normalization factor.

Equation (1.6.2) is oftentimes abbreviated as the product of diagonal elements of the matrix

$$\Psi(1,2,\cdots,N) = \left| \psi_1(1)\alpha(1) \ \psi_1(2)\beta(2) \cdots \psi_n(N)\beta(N) \right| \quad (1.6.3)$$

or a more compact expression

$$\Psi(1,2,\cdots,N) = \left| \psi_1(1)\overline{\psi_1}(2) \cdots \overline{\psi_n}(N) \right| \\ = \left| \psi_1 \overline{\psi_1} \cdots \overline{\psi_n} \right| \quad (1.6.4)$$

In Eq. (1.6.4), the spatial function with a bar on top means that the spin part of spin orbital is β, while the function without a bar means that the spin part is α.

In the notation of Eqs. (1.6.3) and (1.6.4), the normalization factor is understood by convention.

With the use of permutation operator $\hat{P}_k$, Eq. (1.6.2) can be expressed as follows:

$$\Psi(1,2,\cdots,N) = \frac{1}{\sqrt{N!}} \sum_k^{N!} (-1)^k \hat{P}_k \{\psi_1(1)\alpha(1)\psi_1(2)\beta(2) \cdots \psi_n(N)\beta(N)\} \quad (1.6.5)$$

where $\hat{P}_k$ represents the exchange of all coordinates of any pair of electrons, k represents the number of permutations in electron pairs. When k is an odd permutation, $(-1)^k$ is negative; when k is even, $(-1)^k$ is positive.

One of the properties of determinant is that the determinant is multiplied by -1 if we exchange any two rows (or columns) of the determinant. For the determinant in Eq. (1.6.2), this is equivalent to that exchanging all coordinates of any pair of electrons leads to an antisymmetric wave function. Another property of determinant is that if any two columns (or rows) are identical in the determinant, the determinant is equal to zero. For the determinant in Eq. (1.6.2), this means that it is impossible for

1.6 Pauli Exclusion Principle and Periodic Table

two electrons to stay at the completely same spin orbital state. This naturally leads to another expression of Pauli exclusion principle: "No two electrons can occupy the same spin orbital. " In fact, this expression is one deduction of Pauli exclusion principle or antisymmetry principle based on Slater determinant.

As described in Sect. 1.4, electron spin orbital is defined as the product of electron spatial orbital and spin function. To an atomic system, the electron spatial orbital and spin function can be characterized by a set of quantum numbers n, l, m, and m_s. The quantum number n, l, m, and m_s is called principle quantum number, azimuthal quantum number, magnetic quantum number, and spin quantum number, respectively. Thus, another more limited expression of Pauli exclusion principle is: "No two electrons in an atom can have identical quantum numbers n, l, m, and m_s." It is not hard to tell that this expression of Pauli exclusion principle is only confined to the atomic system with central symmetry.

Based on orbital, spin concept, Pauli exclusion principle and the lowest energy principle, we can build the electronic structures for all the elements in the periodic table through Aufbau process. The so-called Aufbau process is a process to build the electronic structures for the elements in atomic order starting from hydrogen atom by adding one proton to the nucleus and one electron to the proper subshell at a time. For example, hydrogen has one proton in the nucleus and one electron outside of the nucleus. Based on the lowest energy principle, this electron should fill 1s orbital, thus, the electron configuration of hydrogen is $1s^1$. Adding one proton in the nucleus of hydrogen leads to the next element—helium, and the electron added at the same time should fill 1s orbital based on the lowest energy principle and Pauli exclusion principle, but the spins of two 1s electrons must be opposite. So the electron configuration of helium is $1s^2$. Next, adding one more proton in the nucleus of helium leads to lithium. The electron added simultaneously should fill 2s orbital based on the lowest energy principle and Pauli principle, so the electron configuration of Lithium is $1s^2 2s^1$. By continuing in this way, the electronic structures for all elements in the whole periodic table can be built. Table 1.1 shows electron configurations for the first 36 elements in the periodic table.

In the Aufbau process, both protons and electrons are added, but for a process by which electrons are added or filled in atoms or molecules, it is often called the Aufbau-like process for building atoms.

Based on the electronic structures of elements, we have many useful concepts such as inner electron, outer electron, open shell, closed shell, valence electron, core electron, as well as σ electron and π electron in chemical bonding, etc.

Table 1.1 Electron configurations for elements $Z = 1 - 36$ in the periodic table

Atomic number	Symbol	Electron configuration	Atomic number	Symbol	Electron configuration
1	H	$1s^1$	19	K	$[Ar]4s^1$
2	He	$1s^2$	20	Ca	$[Ar]4s^2$
3	Li	$[He]2s^1$	21	Sc	$[Ar]3d^1 4s^2$
4	Be	$[He]2s^2$	22	Ti	$[Ar]3d^2 4s^2$
5	B	$[He]2s^2 2P^1$	23	V	$[Ar]3d^3 4s^2$
6	C	$[He]2s^2 2P^2$	24	Cr	$[Ar]3d^5 4s^1$
7	N	$[He]2s^2 2P^3$	25	Mn	$[Ar]3d^5 4s^2$
8	O	$[He]2s^2 2P^4$	26	Fe	$[Ar]3d^6 4s^2$
9	F	$[He]2s^2 2P^5$	27	Co	$[Ar]3d^7 4s^2$
10	Ne	$[He]2s^2 2P^6$	28	Ni	$[Ar]3d^8 4s^2$
11	Na	$[Ne]3s^1$	29	Cu	$[Ar]3d^{10} 4s^1$
12	Mg	$[Ne]3s^2$	30	Zn	$[Ar]3d^{10} 4s^2$
13	Al	$[Ne]3s^2 3P^1$	31	Ga	$[Ar]3d^{10} 4s^2 4P^1$
14	Si	$[Ne]3s^2 3P^2$	32	Ge	$[Ar]3d^{10} 4s^2 4P^2$
15	P	$[Ne]3s^2 3P^3$	33	As	$[Ar]3d^{10} 4s^2 4P^3$
16	S	$[Ne]3s^2 3P^4$	34	Se	$[Ar]3d^{10} 4s^2 4P^4$
17	Cl	$[Ne]3s^2 3P^5$	35	Br	$[Ar]3d^{10} 4s^2 4p^5$
18	Ar	$[Ne]3s^2 3P^6$	36	Kr	$[Ar]3d^{10} 4s^2 4p^6$

1.7 One of the Approximation Methods in Quantum Mechanics—The Variation Method

The atomic or molecular system with two or more electrons is an interacting many-particle system in quantum mechanics. Currently the Schordinger equation of this system can't be solved exactly and can only be treated approximately. The most frequently used approximation method is the variation method and perturbation theory. The following gives a brief introduction related to the variation method.

In quantum chemistry, the variation method has been used in three different ways, i.e. Ritz method, the linear variation method, and specific factor method. They are all based on variation principle.

For an atomic or molecular system with N electrons, the Hamiltonian operator is denoted by $\hat{H}$, and the corresponding Schrodinger equation is

$$\hat{H}\Psi_i = E_i \Psi_i \tag{1.7.1}$$

where Ψ_i and E_i are the eigenfunction and energy eigenvalue for the bound state of the system. $\int \Psi_i^* \Psi_j d\tau = \delta_{ij}$, and $E_0 \leq E_1 \leq \cdots \leq E_i \leq \cdots$ ($i = 0, 1, 2, \cdots$).

1.7 One of the Approximation Methods in Quantum Mechanics—The ...

Although it is impossible to get E_i and Ψ_i by solving Eq. (1.7.1) exactly, an approximate wave function and energy which are close to true solutions can be obtained by the variation principle.

Any function which is continuous, single-valued and quadratic integrable can be called Well-behaved function. The variation principle claims that if we use Well-behaved function Φ which satisfies boundary conditions of the system to replace Ψ, and the expectation value of the operator $\hat{H}$ is W, there must be

$$W = \frac{\Phi|\hat{H}|\Phi}{\Phi|\Phi} \geq E_0 \tag{1.7.2}$$

In the above equation, E_0 is the exact ground state energy, Φ is called the trial wave function, and the approximate ground state energy W is one of the upper bounds of ground state energy E_0.

(1) **Ritz method** [3, 4, 10, 12]

If we choose a trial wave function Φ which contains a number of undetermined parameters

$$\Phi = \Phi(q, c_1, c_2, \cdots, c_i, \cdots) \tag{1.7.3}$$

where q represents all the coordinates of the system, and the rests are undetermined parameters. W obtained by such Φ is a function of undetermined parameter c_i. Based on the variation principle, $W(c_1, c_2, \cdots c_i, \cdots)$ is maximized or minimized,

$$\sum_i \frac{\partial W}{\partial c_i} \delta c_i = 0 \tag{1.7.4}$$

Hereby every parameter c_i and their corresponding W (W_0) can be determined. W_0 is an approximate ground state energy which is closest to E_0 when the trial wave function has the expression in Eq. (1.7.3). The closer the function Φ is to a true wave function, and the more the undetermined parameters c_i are, the smaller W_0 is deviated from E_0.

(2) **Linear variation method**

The trial wave function can be expressed as a linear combination of n linearly independent functions $f_1, f_2, \cdots, f_i, \cdots, f_n$ that satisfy the boundary conditions of the problem,

$$\Phi = \sum_i c_i f_i \tag{1.7.5}$$

Then the undetermined parameters c_i can be determined by variational integral under the conditions of minimization and normalization, and finally the upper bound of energies for the first n states can be obtained. This method is called linear variation method. It is very useful in quantum chemistry calculation.

The details of linear variation method are described as follows:

Rewrite Eq. (1.7.5)

$$\Phi = \sum_i c_i f_i$$

Then variational integral becomes

$$W = \frac{\langle \Phi | \hat{H} | \Phi \rangle}{\langle \Phi | \Phi \rangle} = \frac{\left\langle \sum_i c_i f_i | \hat{H} | \sum_i c_i f_i \right\rangle}{\left\langle \sum_i c_i f_i | \sum_i c_i f_i \right\rangle}$$

$$= \frac{\sum_i \sum_j c_i c_j H_{ij}}{\sum_i \sum_j c_i c_j S_{ij}} \quad (1.7.6)$$

where

$$H_{ij} = f_i | \hat{H} | f_j \quad (1.7.7)$$

$$S_{ij} = f_i | f_j \quad (1.7.8)$$

From Eq. (1.7.6), we can get

$$W \sum_i \sum_j c_i c_j S_{ij} = \sum_i \sum_j c_i c_j H_{ij} \quad (1.7.9)$$

Partial differentiating both sides of the above equation with respect to c_k gives,

$$\frac{\partial W}{\partial c_k} \sum_i \sum_j c_i c_j S_{ij} + W \frac{\partial}{\partial c_k} \sum_i \sum_j c_i c_j S_{ij} = \frac{\partial}{\partial c_k} \sum_i \sum_j c_i c_j H_{ij} \quad (1.7.10)$$

Based on variation principle, to minimize W, requiring

$$\frac{\partial W}{\partial c_k} = 0 (k = 1, 2, \cdots, i, \cdots, j, \cdots, n) \quad (1.7.11)$$

Thus, Eq. (1.7.10) becomes

$$W \frac{\partial}{\partial c_k} \sum_i \sum_j c_i c_j S_{ij} = \frac{\partial}{\partial c_k} \sum_i \sum_j c_i c_j H_{ij} \quad (1.7.12)$$

Then

$$W \sum_i c_i S_{ik} = \sum_i c_i H_{ik} \qquad (1.7.13)$$

or it can be written as

$$\sum_i (H_{ik} - W S_{ik}) c_i = 0 \quad (k = 1, 2, \cdots, n) \qquad (1.7.14)$$

The above expression is homogeneous linear equations with n unknowns $c_1, c_2, \cdots, c_i, \cdots, c_n$, and to get non-zero solutions, the determinant of the secular matrix must be equal to zero, i.e.

$$det(H_{ik} - W S_{ik}) = 0 \qquad (1.7.15)$$

or

$$\begin{vmatrix} H_{11} - S_{11}W & H_{12} - S_{12}W & \cdots & H_{1n} - S_{1n}W \\ H_{21} - S_{21}W & H_{22} - S_{22}W & \cdots & H_{2n} - S_{2n}W \\ \vdots & \vdots & & \vdots \\ H_{n1} - S_{n1}W & H_{n2} - S_{n2}W & \cdots & H_{nn} - S_{nn}W \end{vmatrix} = 0 \qquad (1.7.16)$$

Equations (1.7.15) and (1.7.16) are called secular equations. This equation has n real roots, which are respectively the upper bounds of intrinsic energy levels for the ground state and excited states. The wave functions $\Phi_1, \Phi_2, \cdots, \Phi_n$ corresponding to $W_0, W_1, \cdots, W_n$ are the approximate wave functions for the ground state and excited states of the system.

Now the secular equations can be solved quickly on computer using a standard program, which is one of the great advantages.

(3) The method of Lagrange multipliers

The method of Lagrange multipliers or specific factor method is a conditional variation problem. This method can be used when one wishes the variation of one integral is equal to zero subject to the condition that another integral (or other integrals) must be fixed. The key point of the method is to first write a linear combination of all related integrals I containing Lagrange multiplier (or undetermined factor) λ

$$I_0 + \lambda_1 I_1 + \lambda_2 I_2 + \cdots + \lambda_k I_k \qquad (1.7.17)$$

Then we make the variation of the linear combination equal to zero, i.e.

$$\delta(I_0 + \lambda_1 I_1 + \lambda_2 I_2 + \cdots + \lambda_k I_k) = 0 \qquad (1.7.18)$$

For any given changes, as long as $\delta I_i = 0 (i = 1, 2, \cdots, k)$, there must be $\delta I_0 = 0$. $\lambda_i (i = 1, 2, \cdots, k)$ in above two equations is called Lagrange multiplier [13].

For example:

$$\delta\{\Phi|\hat{A}|\Phi - \lambda[\Phi|\Phi - 1]\} = 0 \qquad (1.7.19)$$

(The proviso is to maintain normalization, i.e. $\Phi|\Phi = 1$) [14, 15]

$$\delta\left[\hat{H} - \sum_{i<j}\sum_{j}\delta(m_{si}, m_{sj})\lambda_{ij}\int\psi_i^*(1)\psi_j(1)d\tau_1\right] = 0 \qquad (1.7.20)$$

(The proviso is to keep the spatial functions mutually orthogonal for single particles with the same spins)

$$\delta\left\{\hat{H} + \sum_{i,j}\lambda_{ij}\left(\delta_{ij} - \int\phi_1^*(1)\phi_j(1)d\tau_1\right)\right\} = 0 \qquad (1.7.21)$$

(The proviso is to maintain orthonormalization, i.e. $\int\phi_1^*(1)\phi_j(1)d\tau_1 = \delta_{ij}$) [15]

The method of Lagrange multipliers is a mathematical technique. It makes it possible for people to add one or more extra provisos (or boundary conditions such as normalization, orthonality, etc.) into variation. If one wants to get the best result by "self-consistent", this method can be useful. About detailed applications of the method of Lagrange multipliers, readers can refer to related references.

References

1. Chu S (1979) Atomic physics. People's Education Press, Beijing
2. Zeng J (2001) Quantum mechanics, vol 1, 3rd edn. Science Press, Beijing
3. Xu G, Li L (1984) Basic principle and ab initio method of quantum chemistry, vol 1. Science Press, Beijing
4. Yin H (1999) Quantum mechanics. University of Science and Technology of China Press, Hefei
5. Levine IN (2005) Quantum chemistry, 5th edn. Prentice-Hall Inc., New Jersey
6. Peng H, Xu X (1998) The basics of theoretical physics. Peiking University Press, Beijing
7. Zheng L, Xu G (1988) Atomic structure and atomic spectroscopy. Peiking University Press, Beijing
8. Xu K (2000) Advanced atomic and molecular physics. Science Press, Beijing
9. Xu G, Li L, Wang D (1985) Basic principle and ab initio method of quantum chemistry, vol 2. Science Press, Beijing
10. Pople JA, Beveridge DL (1976) Approximate molecular orbital theory (trans: Jiang Y). Science Press, Beijing
11. Ballentine LE (2001) Quantum mechanics: a modern development. World Scientific Publishing Co., Pte. Ltd., Singapore
12. Shen Y, Liang Z, Xu L et al (1999) Practical mathematical handbook. Science Press, Beijing
13. Slater JC (1960) Quantum theory of atomic structure, vol I, II. McGraw-Hill Book Company, Inc, New York

14. Springborg M (2000) Methods of electronic-structure calculations: form molecules to solids. John Wiley & Sons Ltd., New York
15. Li J (1984) X_α method and its application in quantum chemistry. Anhui Science and Technology Press, Hefei

Chapter 2
The Weakest Bound Electron Theory (1)

In quantum chemistry, some theories are based on the inidstinguishability of interacting identical particles and antisymmetry principle, such as Hartree-Fock Self-Consistent-Field Method, etc.; while other theories come from the consideration of electronic separability, such as model potential theory for the separation of valence electrons and core electrons, etc. But electrons in an atom or molecule behave with wave-particle duality. Thus, the former shows its insufficient electronic character, and the latter oftentimes needs theoretical support for its separability. How to incorporate the wave-like properties and particular nature of electrons in the same theory, or how to construct a wave-particle unified quantum theory which not only maintain exchange antisymmetry of fermion system and the use of wave function to describe wavelike properties of electrons, but also give consideration to particular nature shown in physics and chemistry that electrons are separable, is what I have always been thinking about and the direction of my research work. We introduced the concept of the Weakest Bound Electron Theory into quantum mechanics and try to construct a new quantum theory based on wave-particle duality.

2.1 The Concept of the Weakest Bound Electron

The concepts such as valence electron, core electron, outer electron, inner electron, frontier electron, non-frontier electron, σ electron and π electron, have been widely accepted by chemists and physicists and frequently applied in quantum theories. By contrast, the concept of the weakest bound electron has less attention, much less becomes the core concept to construct a quantum theory. In fact, this is caused by insufficient estimation of its importance and universality due to the lack of exhaustive study of the weakest bound electron problem. Here, we first introduce the concept of the weakest bound electron into quantum mechanics.

The phenomenon of successive ionization exists in both atoms and molecules. So people are familiar with the definition of ionization energy for atoms or molecules [1, 2] and the process of successive ionization. In Ref. [1], the definition of ionization energy for a free particle is described as: for a free particle, such as an atom or a molecule, the ionization energy is defined as the energy needed to remove the weakest bound electron completely from the particle at ground state so that the generated (positive) ion is also at ground state. Therefore, the energy needed for successive ionization of a neutral particle is called the first ionization energy, the second ionization energy, the third ionization energy, etc., respectively.

This definition gave rise to the concept of the Weakest Bound Electron (or WBE) and Non-Weakest Bound Electron (or N-WBE). The so called Weakest Bound Electron is the electron that is most weakly associated or most loosely bounded with the given system. Thus, it is the electron which can be excited or ionized most easily, and also it is the most active electron in chemistry. Non-Weakest Bound Electron is the electron which associates more strongly or more tightly with the given system and can't be ionized when ionization occurs in the given system. For the system with upcoming ionization process, there is only one Weakest Bound Electron, but the number of Non-Weakest Bound Electron depends on the total number of electrons in the given system. If the total number of electrons is 1, i.e., single electron system, there is only one Weakest Bound Electron and no Non-Weakest Bound Electrons; if the total number of electrons of the given system is N, the system has only one Weakest Bound Electron, while the number of Non-Weakest Bound Electron is $N - 1$.

According to this definition, the only electron of the single electron system (such as hydrogen atom, hydrogen-like ion, and hydrogen molecule ion) is the Weakest Bound Electron of the given system; the valence electron which is outside of a closed shell of alkali metal atom or alkali-like ions is also the Weakest Bound Electron of the given system; the frontier electron in the non-degenerate Highest Occupied Molecular Orbital (HOMO) is the Weakest Bound Electron of the given system as well; the electron at the highest excited energy level of an excited atomic or molecular system is also the Weakest Bound Electron of the given system. Even for an atomic or atomic ion system with more valence electrons as well as a molecular or molecular ion system with two or two more equivalent electrons occupying the highest orbital, these valence electrons will also become the Weakest Bound Electron in sequence which will be ionized, excited or participate in chemical reactions during the process of successive ionization, electron excitation (non inner electron excitation) or chemical reactions. Thus, the Weakest Bound Electron is ubiquitous, and it merely is not given a consensus name of "the Weakest Bound Electron".

In addition, the treatment of electron in a single electron system (i.e. WBE) is very important to the construction of quantum theory; The treatment of the only valence electron of alkali metal atom or alkali-like ions plays an important role in the development of atomic spectroscopy and the construction of quantum defect theory; The concepts such as "HOMO energy level is equal to ionization energy with a changed sign, while LUMO energy corresponds to, or more precisely 'go hand in hand with', electronic affinity with a changed sign" [3], the frontier electron theory, HOMO-LUMO interaction, etc., are significant for explaining organic reaction, stereoselective phenomenon, molecular structure and static properties; The behavior of excited electrons are directly related to energy level transitions, transition state, etc.; Thus, the concept and behavior of the Weakest Bound Electron is so important to natural sciences such as physics, chemistry, etc.

So why has the concept of Weakest Bound Electron been lack of attention in quantum theory? Probably it is due to the following reasons: ① It was thought that there is only one Weakest Bound Electron in an atomic or molecular system with N electrons, so it is impossible to deal with all the problems with N electrons using the concept of Weakest Bound Electron. Actually this is a kind of misunderstanding. ② No proper method has been found to make the Hamiltonian of N electrons to be the sum of Hamiltonian of single electrons. ③ It is associated with the problem of how to express the indistinguishability of identical particles and separability of electrons.

2.2 Ionization Process and Aufbau-Like Process is Reversible

The successive ionization process for a neutral atomic or molecular system with N electrons is shown in Fig. 2.1. In this figure, the N-electron atom (molecule) A is called original system A. During the successive ionization of original system A, electrons are removed one by one as the Weakest Bound Electron. So the successive ionization of original system A can be expressed by a series of subsystems, and the subsystems $A^1, A^2, \ldots, A^N$, represent the neutral N-electron atom (molecule), atomic (molecular) ion with one positive charge, ..., atomic (molecular) ion with $N-1$ positive charges, respectively. During the process of successive ionization, the species which is ionized at every step is the species represented by these subsystems. The subsystem A^1 has only one Weakest Bound Electron and is labeled as WBE^1 here, and at the same time there are $(N-1)$ Non-Weakest Bound Electrons that can't be ionized. These electrons form an aggregate, which is called $core_1$, together with the core or nuclear skeleton of the atom. The subsystem A^2 also has only one Weakest Bound Electron labeled as WBE^2, and $(N-2)$ Non-Weakest Bound Electrons, which form $core_2$ together with the core or nuclear skeleton of the atom. Likewise, the subsystem A^μ also has only one Weakest Bound Electron and $(N-\mu)$ Non-Weakest Bound Electrons which form a core labeled as $core_\mu$. The last subsystem is A^N which also has only one Weakest Bound Electron, but has no Non-Weakest Bound Electrons except for a naked core or naked nuclear skeleton of the atom, which is called $core_N$. Thus, every subsystem has only one Weakest Bound Electron and it is composed of one Weakest Bound Electron and the corresponding core. One subsystem becomes one link of the chain for successive ionization.

The successive ionization chain of the original system A starts from subsystem A^1. After removing WBE_1 from subsystem A^1, the remaining $core_1$ will become subsystem A^2, noticing that "become" is used here, rather than $core_1$ "will be" or "equals" subsystem A^2. If "will be" or "equals" is used, it means that "condition

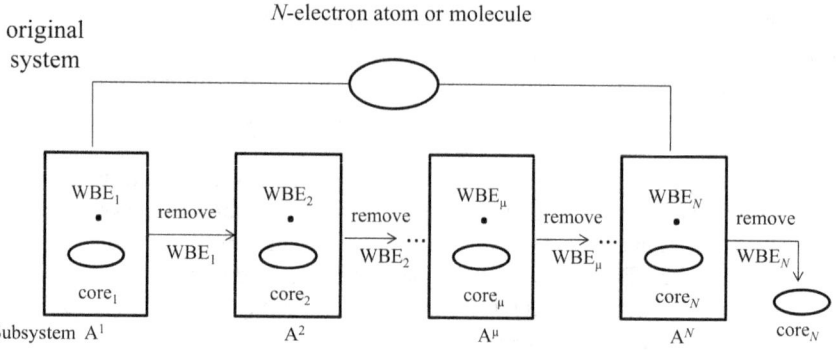

Fig. 2.1 Demonstration of successive ionization for an N-electron atom or molecule. The N electrons in an N-electron atom or molecule are ionized one by one as the weakest bound electron

for frozen orbital" is used. This is the image given by Koopmans theorem and is not what we want to give. There are two meanings for the word "become" that we used: the first is that after removing WBE_1, due to relaxation effect [4, 5], the single electron states of $(N-1)$ Non-Weakest Bound Electrons in $core_1$ will adjust and become different; the second is that for the remaining $(N-1)$ Non-Weakest Bound Electrons in $core_1$, one electron will become the Weakest Bound Electron of subsystem A^2 after adjustment, and the remaining $(N-2)$ electrons will become the Non-Weakest Bound Electrons of subsystem A^2 and form $core_2$ with the core or nuclear skeleton of the atom. Subsequently, subsystems A^2,..., A^μ, ...repeat the same scenario as that for subsystem A^1 until subsystem A^N during the successive ionization. For A^N, after removing WBE_N, only $core_N$ remains and it is a naked core or naked nuclear skeleton. The above is a description of the successive ionization for an atomic or molecular system. After the successive ionization is completed, all electrons ionized as the Weakest Bound Electrons and the core or nuclear skeleton of the atom are infinitely far away or static to each other. The state of the present system is zero energy state of the given system in quantum chemistry. Thus, we have the following relation:

$$E_{electron} = -\sum_{\mu=1}^{N} I_\mu \qquad (2.2.1)$$

where $E_{electron}$ is the total energy of electrons for an N-electron atomic (molecular) system, I_μ represents successive ionization energies. The meaning of this relation is that the total energy of electrons of an N-electron atomic (molecular) system is equal to the negative value of the sum of successive ionization energies.

Next, we will describe how to build an atom or molecule with N-electrons using a process similar to Aufbau process [6, 7] which is used to build an atom in atomic structures. This process is to drag electrons one by one from infinitely far away to

2.2 Ionization Process and Aufbau-Like Process is Reversible

the atomic or molecular environment starting from zero energy state of the system in quantum chemistry. It is a reversible process of successive ionization.

First, we briefly restate the so-called Aufbau process.

To describe the systematic change of the electronic structures for all elements in the periodic table, chemists proposed the so-called Aufbau process, constructing electronic structures for all elements in the periodic table in atomic order. The so-called Aufbau process is to add one proton to the nucleus and one electron to the proper subshell at a time until correct electron configurations of all elements have been constructed starting from hydrogen atom. Specifically, hydrogen atom is composed of an atomic nucleus in which there is one proton with one unit positive charge and one electron outside of the nucleus which is at 1s orbital with the lowest energy state. The electron configuration of hydrogen is $1s^1$. If we add one proton in the nucleus of hydrogen and one electron to 1s orbital with the lowest energy state outside of the nucleus, we can build the second element helium in the periodic table with $1s^2$ electron configuration. If we add one more proton in the nucleus of helium and one more electron to the empty 2s orbital with lowest energy state outside of the nucleus, we can build the third element lithium in the periodic table with $1s^2 2s^1$ electron configuration. By continuing in this way, the electron configurations for all elements in the whole periodic table can be built based on the lowest energy principle and Pauli exclusion principle.

The so-called Aufbau-like process for building atoms in atomic structures is a process of adding electrons one by one in the atomic or molecular environment without adding protons. Let's start from zero energy state of a given system in quantum chemistry, i.e., start from a state in which $core_N$ from Fig. 2.1 is infinitely far away from N electrons which are infinitely far away from each other and they are all static. First, we move one electron from infinitely far away into the surroundings of $core_N$ to construct subsystem A^N; then moving one more electron from infinitely far away into the surroundings of A^N to construct subsystem A^{N-1}. Due to the relaxation effect caused by addition of the second electron, neither of the single electron states of two electrons in A^{N-1} will be the same as the single electron state of that electron in A^N; Thus, through the Aufbau-like process for building atoms in atomic structures, the original system A, i.e. an N-electron atomic or molecular system, can be constructed eventually by constantly adding electrons one by one from infinitely far away to atomic or molecular ions. Due to overlapping energy levels, it is possible that there is no perfect one-to-one match between every step of ionization and every step of adding electrons, but the whole process of adding electrons one by one is a reversible process of the whole successive ionization. Every step of adding electron is the same as every step of ionization with the existence of relaxation effect. Starting from the original system A, the system eventually reaches zero energy state of the given system in quantum chemistry by successive ionization; then, starting from zero energy state, it eventually returns to the state of original system A by adding electrons one by one into the atomic or molecular environment through a Aufbau-like process, and a complete cycle is formed. Thus, for the process of adding electrons, Eq. (2.2.1) can be used as well.

Through above analysis, we get four important results as follows:

(1) According to WBE theory, the N electrons of an N-electron atomic or molecular system can be treated as N Weakest Bound Electrons in N subsystems.
(2) Since removing electrons (successive ionization) and adding electrons (Aufbau-like process) are reversible and they forms a closed cycle, for N electrons in an N-electron atomic or molecular system, treating them as N Weakest Bound Electrons in N subsystems is equal to treating them as a combination of electron 1, electron 2, ..., electron N in original system A. Although there seems a big difference between two images and different treatments in front of us, in fact, by nature, the former just renames electron 1, electron 2, ..., electron N using the name of the Weakest Bound Electron based on the idea of dynamic ionization. According to the content of quantum mechanics described in Chap. 1, if electrons are simply renamed or renumbered, the Hamiltonian of the system keeps constant. [8, 9]
(3) In subsystem A^μ, the Weakest Bound Electron μ is moving in a potential field of core$_\mu$ composed of (N-μ) Non-Weakest Bound Electrons and the core or nuclear skeleton of the atom. (Noticing that core$_N$ is only an atomic core or nuclear skeleton)
(4) The state in which electron and electron, or electron and atomic core or constant nuclear skeleton, are infinitely far way and static for the given system in quantum chemistry is zero energy state of the system (or the standard state of quantum chemistry), therefore, Eq. (2.2.1) is always true for either treatment of removing or adding electrons. Here there is no Koopmans theorem approximation.

2.3 The One-Electron Hamiltonian for the Weakest Bound Electron

2.3.1 The Non-Relativistic One-Electron Hamiltonian for the Weakest Bound Electron

The non-relativistic electronic Hamiltonian (atomic unit) of an N-electron atom or molecule is

$$\hat{H}^{nr} = \sum_{\mu=1}^{N}\left(-\frac{1}{2}\nabla_\mu^2\right) - \sum_{A=1}^{X}\sum_{\mu=1}^{N} Z_A r_{A\mu}^{-1} + \sum_{\mu<\upsilon}\sum_{\upsilon}^{N} r_{\mu\upsilon}^{-1} \qquad (2.3.1)$$

Since the Hamiltonian of the system doesn't not change when electrons are renamed using the name of the Weakest Bound Electron, the non-relativistic Hamiltonian of the system is still Eq. (2.3.1) when treating N electrons in an N-electron atomic or molecular system as N Weakest Bound Electrons.

2.3 The One-Electron Hamiltonian for the Weakest Bound Electron

We examine Eq. (2.3.1) from a dynamic ionization standpoint. After re-assembling all terms on the right side of Eq. (2.3.1) based on the opinion of dynamic ionization, this equation becomes

$$\hat{H}^{nr} = \left(-\frac{1}{2}\nabla_1^2 - \sum_{A=1}^{X} Z_A r_{A1}^{-1} + \sum_{v,1<v}^{N} r_{1v}^{-1} \right)$$

$$+ \left(-\frac{1}{2}\nabla_2^2 - \sum_{A=1}^{X} Z_A r_{A2}^{-1} + \sum_{v,2<v}^{N} r_{2v}^{-1} \right)$$

$$+ \left(-\frac{1}{2}\nabla_\mu^2 - \sum_{A=1}^{X} Z_A r_{A\mu}^{-1} + \sum_{v,\mu<v}^{N} r_{\mu v}^{-1} \right)$$

$$+ \left(-\frac{1}{2}\nabla_N^2 - \sum_{A=1}^{X} Z_A r_{AN}^{-1} \right) \quad (2.3.2)$$

In the first stage of ionization, after electron 1 was removed as WBE$_1$, all terms in the first parenthesis on the right side of Eq. (2.3.2) disappeared, because these terms are related to electron 1, i.e., removed WBE$_1$. Then, in the second stage of ionization, after electron 2 was removed as WBE$_2$, accordingly all terms in the second parenthesis disappeared since these terms are related to electron 2, i.e., removed WBE$_2$. And so on until the N stage of ionization. In the N stage of ionization, when the last electron, i.e., electron N, was removed as WBE$_N$, all terms in the last parenthesis on the right side of Eq. (2.3.2) disappeared. The state of the system is zero energy state in quantum chemistry that was described before. Thus, the non-relativistic Hamiltonian of an N-electron system can be written as a sum of non-relativistic one-electron Hamiltonians of N Weakest Bound Electrons based on WBE theory, i.e.

$$\hat{H}^{nr} = \sum_{\mu=1}^{N}\left(-\frac{1}{2}\nabla_\mu^2\right) - \sum_{A=1}^{X}\sum_{\mu=1}^{N} Z_A r_{A\mu}^{-1} + \sum_{\mu<v}^{N}\sum_{v}^{N} r_{\mu v}^{-1}$$

$$= \sum_{\mu=1}^{N}\left(-\frac{1}{2}\nabla_\mu^2 - \sum_{A=1}^{X} Z_A r_{A\mu}^{-1} + \sum_{v,\mu<v}^{N} r_{\mu v}^{-1} \right)$$

$$= \sum_{\mu=1}^{N} \hat{H}_\mu^{nr} \quad (2.3.3)$$

Then we have

$$\hat{H}_\mu^{nr} = -\frac{1}{2}\nabla_\mu^2 - \sum_{A=1}^{X} Z_A r_{A\mu}^{-1} + \sum_{\nu,\mu<\nu}^{N} r_{\mu\nu}^{-1}$$
$$= \hat{h}(\mu) + \sum_{\nu,\mu<\nu}^{N} r_{\mu\nu}^{-1} = \hat{h}(\mu) + \sum_{\nu,\mu<\nu}^{N} r_{\mu\nu}^{-1} \quad (2.3.4)$$

where

$$\hat{h}(\mu) = -\frac{1}{2}\nabla_\mu^2 - \sum_{A=1}^{X} Z_A r_{A\mu}^{-1} \quad (2.3.5)$$

$\hat{H}_\mu^{nr}$ is the non-relativistic one-electron Hamiltonian for WBE$_\mu$. $\hat{h}(\mu)$ represents the sum of kinetic energy operator for WBE$_\mu$ and the attractive potential operators between WBE$_\mu$ and all atomic cores in the system. $\sum_{\nu,\mu<\nu}^{N} r_{\mu\nu}^{-1}$ represents the sum of the repulsive potential operators between WBE$_\mu$ and Non-Weakest Bound Electron ν in the subsystem A$^\mu$.

Now, we examine the change of Hamiltonian with addition of electrons through a reversible process of successive ionization, i.e., Aufbau-like process by adding electrons.

Starting from zero energy state, a ground-state atomic or molecular ion with one electron can be built by moving an electron which is infinitely far away and static (randomly label it as electron N) into the atomic or molecular ion environment of a naked core or naked nucleus skeleton. Hamiltonian for this ion is

$$\hat{H}^{nr}(N) = -\frac{1}{2}\nabla_N^2 - \sum_{A=1}^{X} Z_A r_{AN}^{-1} \quad (2.3.6)$$

For this ion, apparently the introduced electron N is exactly the Weakest Bound Electron of the constructed system. Using the same label as for successive ionization, this ion with $(N-1)$ positive charges is subsystem A^N. The only electron is WBE$_N$, so

$$\hat{H}^{nr}(N) = -\frac{1}{2}\nabla_N^2 - \sum_{A=1}^{X} Z_A r_{AN}^{-1} = \hat{H}_{WBE_N}^{nr} \quad (2.3.7)$$

Then, by moving one more electron which is infinitely far away and static [randomly label it as electron $(N-1)$] into the ionic environment with $(N-1)$ positive charges, a ground-state atomic or molecular ion with two electrons can be built. Hamiltonian of this system is

2.3 The One-Electron Hamiltonian for the Weakest Bound Electron

$$\hat{H}^{nr}(N-1, N) = -\frac{1}{2}\nabla_{N-1}^2 - \sum_{A=1}^{X} Z_A r_{A(N-1)}^{-1} + r_{(N-1)N}^{-1}$$

$$+ \left(-\frac{1}{2}\nabla_N^2 - \sum_{A=1}^{X} Z_A r_{AN}^{-1} \right) \quad (2.3.8)$$

Comparing Eq. (2.3.8) with Eq. (2.3.6), we see that three more terms are added to the right side of Eq. (2.3.8), i.e., the kinetic energy operator of electron $(N-1)$, the attractive potential operator between electron $(N-1)$ and all the atomic cores, and the repulsive potential operator between electron $(N-1)$ and electron N. By continuing in this way, a ground-state neutral atom or molecule with N electrons, i.e., the original system A, can be built by moving the last electron which is infinitely far away and static (randomly label it as electron 1) into an atomic or molecular ion with one positive charge. Thus, Hamiltonian of the system is

$$\hat{H}^{nr}(1, 2, \cdots, N) = \left(-\frac{1}{2}\nabla_1^2 - \sum_{A=1}^{X} Z_A r_{A1}^{-1} + \sum_{v, 1 < v}^{N} r_{1v}^{-1} \right)$$

$$+ \left[\sum_{\mu=2}^{N} \left(-\frac{1}{2}\nabla_\mu^2 \right) - \sum_{A=1}^{X}\sum_{\mu=2}^{N} Z_A r_{A\mu}^{-1} + \sum_{\mu=2}^{N}\sum_{\mu<v}^{N} r_{\mu v}^{-1} \right]$$

$$= \sum_{\mu=1}^{N} \left(-\frac{1}{2}\nabla_\mu^2 \right) - \sum_{A=1}^{X}\sum_{\mu}^{N} Z_A r_{A\mu}^{-1} + \sum_{\mu<v}^{N}\sum_{v}^{} r_{\mu v}^{-1} \quad (2.3.9)$$

This equation is the same as Eq. (2.3.1). Starting from zero energy state, with electrons added in succession, eventually we have

$$\hat{H}^{nr}(1, 2, \cdots, N) = \sum_{\mu=1}^{N} \left(-\frac{1}{2}\nabla_\mu^2 \right) - \sum_{A=1}^{X}\sum_{\mu}^{N} Z_A r_{A\mu}^{-1} + \sum_{\mu<v}^{N}\sum_{v}^{} r_{\mu v}^{-1}$$

$$= \sum_{\mu=1}^{N} \hat{H}_\mu^{nr} = \hat{H}^{nr} \quad (2.3.10)$$

The above shows that the processes of removing and adding electrons give the same results. In other words, Hamiltonian of the system does not change when electrons are renamed, and the non-relativistic Hamiltonian of an N-electron system can be written as a sum of non-relativistic one-electron Hamiltonians of N Weakest Bound Electrons.

In quantum mechanics and quantum chemistry, we all hope that Hamiltonian of a many-electron system can be written as a sum of one-electron Hamiltonians. But due to the existence of r_{ij}^{-1}, variables can't be separated, so many approximations are proposed. For instance, in the independent particle approximation, many-electron

Hamiltonian can be written as a sum of one-electron Hamiltonians. In the self-consistent field approximation, one-electron potential function can be regarded as a mean field potential generated by the atomic core and (N-1) electrons through duplicated calculations of r_{ij}^{-1}, so that a modified many-electron Hamiltonian can be written as a sum of "effective" one-electron Hamiltonians. This modified many-electron Hamiltonian is different from the original Hamiltonian of the system after all, so a new concept of orbital energy was defined [9–12]. It is generally known that the problem of hydrogen atom with a single electron can be solved exactly in quantum mechanics. The Schrodinger equation of hydrogen atom can be written as

$$\hat{H}^0 \psi^0 = \varepsilon^0 \psi^0 \qquad (2.3.11)$$

where ψ^0 is the electron wave function, also called orbital; ε^0 is the energy of electron on the orbital. Orbital and electronic energy are interrelated. While in the self-consistent field approximation, the eigenfunction of one-electron operator is called orbital, and the eigenvalue corresponding to the orbital is called orbital energy instead of the energy of electron on the orbital. In other words, orbital is correlated with orbital energy but not the energy of electron on the orbital any more. Orbital energy and electronic energy are approximately equal only under Koopmans approximation [4, 5], and the total electronic energies of the system are not equal to the sum of orbital energies. When the concept of the Weakest Bound Electron and the idea of successive ionization were introduced into quantum mechanics, the non-relativistic many-electron Hamiltonian can be written as the sum of non-relativistic one-electron Hamiltonians of N Weakest Bound Electrons without any approximations.

2.3.2 The Treatment of Magnetic Interaction Between Electrons

All effects in atoms or molecules are mainly associated with electromagnetic interactions. The non-relativistic Hamiltonian of the system includes Coulomb interactions between electron and core as well as between electron and electron, but it doesn't include all magnetic interactions. Taking atomic system as an example, except for Coulomb interactions and magnetic interaction between electron and core, the magnetic interactions related to electron include the self spin-orbital coupling of single electrons and magnetic coupling between two electrons μ and υ(including the coupling of spin of electron μ and spin of electron υ, the coupling of spin of electron μ and orbital of electron υ, the coupling of orbital of electron μ and orbital of electron υ, and coupling of orbital of electron μ and spin of electron υ.) A full Hamiltonian should include all these couplings. The use of the concept of the Weakest Bound Electron and the idea of successive ionization can also precisely find where the magnetic interactions between these electrons belong to. The spin-orbital coupling operator $\hat{H}_{so}$ of single electron, needless to say, of course belongs to one-electron Hamiltonian

2.3 The One-Electron Hamiltonian for the Weakest Bound Electron

of each Weakest Bound Electron. While for all kinds of magnetic coupling operators between two electrons, their ownerships can be classified in a way exactly like treating $r_{\mu\nu}^{-1}$ in the previous subsection. Taking, for example, the coupling operator $\hat{H}_{soo}$ of spin of electron μ and orbital of other electrons υ, the coupling operator of spin of electron 1 treated as WBE_1 and orbital of electron 2, electron 3, ..., electron N should belong to one-electron Hamiltonian of WBE_1; Similarly, the coupling operator of spin of electron 2 treated as WBE_2 and orbital of electron 3, electron 4, ..., electron N should belong to one-electron Hamiltonian of WBE_2. In this way, all magnetic couplings between two electrons can find where they belong to.

2.3.3 Relativistic Hamiltonian

Considering the Weakest Bound Electron μ, when ignoring the magnetic interactions between electron and atomic core as well as all kinds of inter-electron electronic and magnetic interactions between the Weakest Bound Electron μ and (N − μ) Non-Weakest Bound Electrons in subsystem A^μ, the relativistic Hamiltonian of the Weakest Bound Electron μ can have the expression of one-electron Dirac Hamiltonian $H_D(\mu)$. Based on this, if considering the repulsive interactions between the Weakest Bound Electron μ and Non-Weakest Bound Electron υ, a new term, i.e.

$$\sum_{\nu,\mu<\nu}^{N} r_{\mu\nu}^{-1} \qquad (2.3.12)$$

should be included. In the above formula, $r_{\mu\nu}$ is the distance between the Weakest Bound Electron μ and Non-Weakest Bound Electron υ. If considering again the relativistic interactions between the Weakest Bound Electron μ and Non-Weakest Bound Electron υ, including WBE_μ spin—$NWBE_\nu$ spin coupling, WBE_μ spin—$NWBE_\nu$ orbital coupling, WBE_μ orbital—$NWBE_\nu$ spin coupling, WBE_μ orbital—$NWBE_\nu$ orbital coupling, and retardation effect, a new term called Breit operator should be added in. Thus, after ignoring the magnetic interactions between electron and atomic core, the relativistic Hamiltonian of the Weakest Bound Electron can be expressed as (in atomic unit)

$$H_\mu^R = H_D(\mu) + \sum_{\nu,\mu<\nu}^{N} \frac{1}{r_{\mu\nu}} + \sum_{\nu,\mu<\nu}^{N} B(\mu,\nu) \qquad (2.3.13)$$

where

$$H_D(\mu) = \alpha(\mu)[cP(\mu) + A(r_\mu)] + c^2\beta(\mu) - \phi(r_\mu) \qquad (2.3.14)$$

In Eq. (2.3.14), $A(r_\mu)$ and $\phi(r_\mu)$ represents the vector potential and scalar potential, respectively, of the electromagnetic field acted on the Weakest Bound Electron μ. In Pauli-Dirac representation, the matrix of $\alpha(\mu)$ and $\beta(\mu)$ is expressed as follows [13–15]:

$$\alpha(\mu) = \begin{bmatrix} 0 & \sigma^P \\ \sigma^P & 0 \end{bmatrix} \quad (2.3.15)$$

$$\beta(\mu) = \begin{bmatrix} I & 0 \\ 0 & -I \end{bmatrix} \quad (2.3.16)$$

Here, σ^P are three 2 × 2 Pauli matrix, and I is 2X2 unit matrix.

The Breit operator in Eq. (2.3.13) has been traditionally treated as a sum of two following terms, i.e., magnetic interactions

$$g^M(\mu, \nu) = -\frac{\alpha(\mu)\alpha(\nu)}{r_{\mu\nu}} \quad (2.3.17)$$

and retardation term,

$$g^R(\mu, \nu) = \frac{1}{2}\left\{\frac{\alpha(\mu)\alpha(\nu)}{r_{\mu\nu}} - \frac{[\alpha(\mu)r_{\mu\nu}][\alpha(\nu)r_{\mu\nu}]}{r_{\mu\nu}^3}\right\} \quad (2.3.18)$$

After substituting Eq. (2.3.14) into Eq. (2.3.13), under the condition of ignoring Breit interaction term and central field approximation [Now $A(r_\mu) = 0$, $\phi(r_\mu) = Z/r_\mu$], the relativistic Hamiltonian operator H'_μ of the Weakest Bound Electron μ can be written as

$$H'_\mu = \alpha(\mu)cP(\mu) + c^2\beta + V(r_\mu) \quad (2.3.19)$$

where

$$V(r_\mu) \approx -\frac{Z}{r_\mu} + \sum_{\nu,\mu<\nu}^{N} \frac{1}{r_{\mu\nu}} \quad (2.3.20)$$

By a similar treatment as in reference [14, 15], Eq. (2.3.19) can be further written as

$$\begin{aligned} H'_\mu = &\frac{1}{2}[P(\mu)]^2 + V(r_\mu) - \frac{1}{8c^2}[P(\mu)]^4 \\ &+ \frac{1}{2c^2 r_\mu}\frac{dV(r_\mu)}{dr_\mu}(s_\mu l_\mu) - \frac{1}{4c^2}\frac{dV(r_\mu)}{dr_\mu}\frac{\partial}{\partial r_\mu} \end{aligned} \quad (2.3.21)$$

or

2.4 The One-Electron Schrodinger Equation of the Weakest Bound Electron

$$H'_\mu = \frac{1}{2}[P(\mu)]^2 + V(r_\mu) - \frac{1}{8c^2}[P(\mu)]^4$$
$$+ \frac{1}{2c^2 r_\mu}\frac{dV(r_\mu)}{dr_\mu}(s_\mu l_\mu) + \frac{1}{8c^2}\nabla^2 V(r_\mu) \quad (2.3.22)$$

The first two terms in Eq. (2.3.21) or Eq. (2.3.22) are actually the non-relativistic Hamiltonian H^0_μ of the Weakest Bound Electron μ under central field approximation, the third term is mass-velocity term denoted by $\Delta \hat{H}_m$, the fourth term is the self spin-orbital coupling of the Weakest Bound Electron μ denoted by $\Delta \hat{H}_{ls}$, and the last term represents the Darwin term denoted by $\Delta \hat{H}_d$.

From this section, we can that the relativistic or non-relativistic Hamiltonian of an N-electron atomic or molecular system is equal to the sum of one-electron relativistic or non-relativistic Hamiltonians of N Weakest Bound Electron, i.e., additivity, based on WBE theory.

2.4 The One-Electron Schrodinger Equation of the Weakest Bound Electron

Below is the discussion about atomic system.

For subsystem A^μ, let $\hat{H}_\mu$ represent one-electron Hamiltonian of WBEμ. ψ_μ and ε_μ represent its wave function and energy, respectively. Thus, the one-electron Schrodinger equation of WBEμ is

$$\hat{H}_\mu \psi_\mu = \varepsilon_\mu \psi_\mu \quad (2.4.1)$$

And we have

$$\langle \psi_\mu | \hat{H}_\mu | \psi_\mu \rangle = \varepsilon_\mu = -I_\mu \quad (2.4.2)$$

For N Weakest Bound Electrons, we have

$$\sum_{\mu=1}^{N} \langle \psi_\mu | \hat{H}_\mu | \psi_\mu \rangle = \sum_{\mu=1}^{N} \varepsilon_\mu = -\sum_{\mu=1}^{N} I_\mu = E_{electron} \quad (2.4.3)$$

In the above equation, I_μ represents μth ionization energy of an N-electron system. $E_{electron}$ represents the total electronic energy of the system.

For a molecular system, under fixed nuclei approximation, the total energy of the system E_{total} is equal to the total electronic energy $E_{electron}$ plus the inter-nucleus repulsive energy [16], i.e.

$$E_{total} = E_{electron} + \sum_{B}^{X} \sum_{A<B}^{X} Z_A Z_B R_{AB}^{-1} \qquad (2.4.4)$$

where R_{AB} is the nuclear distance between nucleus A and nucleus B.

When ignoring the magnetic interactions between electron and nucleus, $\hat{H}_\mu$ can be written in an expression like (2.3.13)

$$\hat{H}_\mu^R = H_D(\mu) + \sum_{\nu,\mu<\nu}^{N} \frac{1}{r_{\mu\nu}} + \sum_{\nu,\mu<\nu}^{N} B(\mu,\nu) \qquad (2.4.5)$$

If we further ignore Breit operator and use central field approximation, $\hat{H}_\mu^R$ will have the expression like (2.3.21) or (2.3.22), i.e.

$$\hat{H}_\mu' = \frac{1}{2}[P(\mu)]^2 + V(r_\mu) - \frac{1}{8c^2}[P(\mu)]^4$$
$$+ \frac{1}{2c^2 r_\mu} \frac{dV(r_\mu)}{dr_\mu}(s_\mu l_\mu) - \frac{1}{4c^2} \frac{dV(r_\mu)}{dr_\mu} \frac{\partial}{\partial r_\mu} \qquad (2.4.6)$$

where

$$V(r) \approx -\frac{Z}{r_\mu} + \sum_{\nu=\mu+1}^{N} \frac{1}{r_{\mu\nu}} \qquad (2.4.7)$$

If we again ignore the last three terms on the right side of Eq. (2.4.6), we will get the expression for the non-relativistic one-electron Hamiltonian of the Weakest Bound Electron under central field approximation,

$$\hat{H}_\mu^0 = -\frac{1}{2}[P(\mu)]^2 + V(r_\mu) = -\frac{1}{2}\nabla_\mu^2 + V(r_\mu) \qquad (2.4.8)$$

The corresponding non-relativistic one-electron Schrodinger equation of the Weakest Bound Electron is

$$\hat{H}_\mu^0 \psi_\mu^0 = \varepsilon_\mu^0 \psi_\mu^0 \qquad (2.4.9)$$

where ε_μ^0 and ψ_μ^0 are the eigenvalue and eigenfunction of $\hat{H}_\mu^0$, respectively.

2.5 The Key Points of the WBE Theory

The key points of the WBE theory have been summed up as follows.

(1) The whole process of successive ionization of an N-electron system by removing the Weakest Bound Electrons and the whole Aufbau-like process by adding electrons one by one to build an N-electron system are mutually reversible. The two processes form a closed cycle. Removing and adding electrons reveal two modes for dealing with problems of N-electron atoms or molecules. The two modes only rename electrons, so the Hamiltonian of the system doesn't change.

The two modes mean that electrons can be handled as a whole in a system, and they also can be handled one by one. This not only indicates indistinguishability of identical particles and antisymmetry principle but also find theoretical foundations for separability of electrons, manifesting the character of electrons.

Both modes have

$$E_{electron} = -\sum_{\mu}^{N} I_\mu \qquad (2.5.1)$$

and also have the same zero state of total electronic energy selected in quantum chemistry.

(2) Every time when removing or adding an electron, relaxation effects are always accompanied, i.e. the change of the number of electrons cause the change of all single-particle states of the system.

$$\varepsilon_\mu = -I_\mu \qquad (2.5.2)$$

References

1. Thewlis J et al (1961) Encyclopedic dictionary of physics, vol 2. Oxford: Pergamon Press
2. Cowan RD (1981) The theory of atomic structure and spectra. University of California Press, Berkeley
3. Fukui K (1985) Chemical reaction and electron orbital (trans: Li R). Science Press, Beijing
4. Veszpremi T, Feher M (1999) Quantum chemistry: fundamentals to applications. Kluwer Academic/Plenum Publishing, New York
5. Springborg M (2000) Methods of electronic-structure calculations: form molecules to solids. John Wiley & Sons Ltd., New York
6. Steinfeld JI (1974) Molecules and radiation: an introduction to modern molecular spectroscopy. Harper & Row, Publishers Inc., New York
7. Holtzclaw HF, Jr., Robinson W R, Odom J D, (1991) General chemistry with qualitative analysis, 9th edn. D.C. Heath and Company, Lexington
8. Ballentine LE (2001) Quantum mechanics: a modern development. World Scientific Publishing Co., Pte. Ltd., Singapore
9. Yin H (1999) Quantum mechanics. University of Science and Technology of China Press, Hefei

10. Pople JA, Beveridge DL (1976) Approximate molecular orbital theory. Translated by Jiang Y. Science Press, Beijing
11. Zeng J (2001) Quantum mechanics, vol 1, 3rd edn. Science Press, Beijing
12. Murrell JN, Kettle SFA, Tedder JM (1978) Valence theory. Translated by Wen Z, Yao W, et al. Science Press, Beijing
13. Peng H, Xu X (1998) The basics of theoretical physics. Peiking University Press, Beijing
14. Zheng L, Xu G (1988) Atomic structure and atomic spectroscop. Peiking University Press, Beijing
15. Zeng J (2001) Quantum mechanics, vol 2. Science Press, Beijing
16. Liao M, Wu G, Liu H (1984) Ab initio quantum chemistry methods. Tsinghua University Press, Beijing
17. Xu G, Li L, Wang D (1985) Basic principle and ab initio method of quantum chemistry, vol 2. Science Press, Beijing
18. Bethe HA, Salpeter EE (1957) Quantum mechanics of one-and two-electron atoms. Springer-Verlag, Berlin
19. Grant IP (1970) Relativistic calculation of atomic structure. Adv Phys 19:747–811
20. Kim YK (1967) Relativistic self-consistent-field theory for closed-shell atoms. Phys Rev 154:17
21. Mann JB, Johnson WR (1971) Breit interaction in multielectron atoms. Phys Rev A 4:41–51
22. Mann JB, Waber JT (1973) Self-consistent relativistic Dirac-Hartree-Fock calculations of lanthanide atoms. At Data Nucl Data Tables 5:201–229
23. Desclaux JP (1973) Relativistic Dirac-Fock expectation values for atoms with $Z = 1$ to $Z = 120$. At Data Nucl Data Tables 12:311–406
24. Adv KFW, At, (1999) High-precision calculations for the ground and excited states of the lithium atom. Mol Opt Phys 40:57–112
25. Zheng NW, Wang T, Ma DX et al (2004) Weakest bound electron potential model theory. Int J Quantum Chem 98:281–290

Chapter 3
The Weakest Bound Electron Theory (2)

From angle [1–3] of particle interactions, there is no difference of quantum mechanical treatment between an atom and a molecule as a many-particle system of fermions [4]. So the basic idea of WBE theory discussed in previous chapter can be applied to both atomic and molecular systems. But since molecules are composed of atoms, they are not at the same level after all and apparently there is difference between them. For instance, atoms and molecules have different symmetry. Atoms are centrosymmetric, but molecules have point-group symmetry. Thus, in this chapter we will discuss atoms in depth and propose an analytical expression of WBE theory in atomic systems—Weakest Bound Electron Potential Model Theory or WBEPM Theory.

3.1 Potential Function

Central-force field problems are important in either classical mechanics or quantum mechanics. Atoms are centrosymmetric, so central-force field is more important with respect to atoms.

The electron of hydrogen atom moves in the Coulomb field of the atomic nucleus, so it is a central-force field problem. The single valence electron of an alkali metal atom moves in a potential field of the atomic core consisting of the stable electronic structure of noble gases and the atomic nucleus. It is also a central-force field problem with very good approximation. The so-called central-force field is that the potential of a moving particle is only a function of the distance between the particle and the center of the force field [6].

The study of the problem of atomic structures using the central-force field model began in the 1920s. In quantum mechanics, both Hartree method and Hartree–Fock self-consistent-field method are also based on the central-force field model.

The spectral characteristics of alkali metal atoms are very similar to those of hydrogen and hydrogen-like atoms, but there are differences between them. The electron of hydrogen and hydrogen-like atoms moves in an electric field generated by a point charge located at the atomic nucleus, but the single valence electron of an alkali atom moves in a potential field generated by a non-rigid atomic core. The atomic core that interacts with the single valence electron is more like a complicated charge system than a point charge. The potential energy of a single valence electron can be expressed by the following series:

$$V = -\frac{e^2}{r} - C_1\frac{e^2}{r^2} - C_2\frac{e^2}{r^3} - \cdots \qquad (3.1.1)$$

where the first term represents the potential energy of the single valence electron in an electric field generated by a shield nucleus with $+Z_{net}e$ charges; the second term represents the potential energy of the single valence electron in a dipole field caused by the polarization of the atomic core due to the single valence electron; the remaining terms represent the potential energies in an electric field of a complicated charge system. The first two terms of Eq. (3.1.1) can explain the spectral characteristics of alkali metal atoms very well [6–8].

In the previous chapter, we have illustrated that when the concept of the Weakest Bound Electron was introduced into quantum mechanics, the problem of an N-electron atom or molecule can treated as the problem of N Weakest Bound Electrons, and we also derived the non-relativistic one-electron Hamiltonian as well as the corresponding one-electron Schrodinger equation of the Weakest Bound Electron for an atomic system under the central field approximation, i.e.

$$\hat{H}_\mu^0 = -\frac{1}{2}\nabla_\mu^2 + V(r_\mu) \qquad (3.1.2)$$

where

$$V(r_\mu) \approx -\frac{z}{r_\mu} + \sum_{\nu=\mu+1}^{N} \frac{1}{r_{\mu\nu}} \qquad (3.1.3)$$

and

$$\hat{H}_\mu^0 \psi_\mu^0 = \varepsilon_\mu^0 \psi_\mu^0 \qquad (3.1.4)$$

How to write an approximate analytical expression of $V(r_\mu)$?

The Weakest Bound Electron μ moves in an electric field of atomic core (or core$_\mu$) consisting of $(N-\mu)$ Non-Weakest Bound Electrons and the atomic nucleus, which is somewhat similar to the motion of a single valence electron of an alkali metal atom. Thus, it should be a good choice to approximate $V(r_\mu)$ using the first two terms of Eq. (3.1.1) which has central field properties. However, more interactions between electrons should be considered.

The Weakest Bound Electron μ is not only attracted to the atomic nucleus but also repelled by $(N-\mu)$ electrons. The repulsion of inner-layer electrons is equivalent to shielding nuclear charge. If the shielding effect is complete, the net nuclear charge

3.2 The Solution of the Radial Equation

experienced by the Weakest Bound Electron μ is $+Z_{net}e$ (for a neutral atom, $Z_{net} = 1$; for an ion with one positive charge, $Z_{net} = 2$; ...). However, "core" is not rigid and the electrons penetrate[7]. Due to the penetration effect, the shielding is not complete. Thus, the Weakest Bound Electron μ should feel a Coulomb field generated by an effective nuclear charge $+Z'e$ instead of net nuclear charge $+Z_{net}e$. So we use $+Z'e$ to replace $+Z_{net}e$ in the first term of Eq. (3.1.1). At the same time, we use B instead of C_1 for the coefficient of the second term in Eq. (3.1.1) in order to differentiate from discussions about alkali metal atoms. Then $V(r_\mu)$ can be expressed approximately as follows [1, 2, 4, 9]:

$$V(r_\mu) \approx -\frac{z'}{r_\mu} + \frac{B}{r_\mu^2} \text{(atomic unit)} \qquad (3.1.5)$$

Substituting (3.1.5) into (3.1.2), we get

$$\hat{H}_\mu^0 = -\frac{1}{2}\nabla_\mu^2 + V(r_\mu) \approx -\frac{1}{2}\nabla_\mu^2 - \frac{z'}{r_\mu} + \frac{B}{r_\mu^2} \qquad (3.1.6)$$

Substituting (3.1.6) into (3.1.4), then

$$\left[-\frac{1}{2}\nabla_\mu^2 - \frac{z'}{r_\mu} + \frac{B}{r_\mu^2}\right]\psi_i''(\mu) = \varepsilon_\mu'' \psi_i''(\mu) \qquad (3.1.7)$$

where $\psi_i''(\mu)$ and ε_μ'' are approximations of ψ_μ^0 and ε_μ^0, respectively, at given analytical potential. In Eq. (3.1.7), orbital is labeled as i and electron is labeled as μ for possible convenience. If all labels are neglected, Eq. (3.1.7) becomes

$$\left[-\frac{1}{2}\nabla^2 - \frac{z'}{r} + \frac{B}{r^2}\right]\psi = \varepsilon\psi \qquad (3.1.8)$$

This equation can be precisely solved and the expression of B can be obtained.

3.2 The Solution of the Radial Equation

3.2.1 Spherical Harmonic

Rewrite Eq. (3.1.8)

$$\left[-\frac{1}{2}\nabla^2 - \frac{z'}{r} + \frac{B}{r^2}\right]\psi = \varepsilon\psi \qquad (3.2.1)$$

Let

$$-Z' = A \tag{3.2.2}$$

and transform Cartesian coordinates into polar coordinates, then Eq. (3.2.1) becomes

$$\frac{1}{2}\left[\frac{1}{r^2}\frac{\partial}{\partial r}\left(r^2\frac{\partial \psi}{\partial r}\right) + \frac{1}{r^2 \sin\theta}\frac{\partial}{\partial \theta}\left(\sin\theta \frac{\partial \psi}{\partial \theta}\right)\right. \\ \left. + \frac{1}{r^2 \sin^2\theta}\frac{\partial^2 \psi}{\partial \phi^2}\right] + \left(\varepsilon - \frac{A}{r} - \frac{B}{r^2}\right)\psi = 0 \tag{3.2.3}$$

where

$$\psi = \psi(r, \theta, \phi)$$

Using separation of variables, we assume

$$\psi(r, \theta, \phi) = R(r)Y(\theta, \phi) \tag{3.2.4}$$

After substituting it into (3.2.3), the equation can be separated into the radial equation and the angular equation

$$-\left\{\frac{1}{Y \sin\theta}\frac{\partial}{\partial \theta}\left(\sin\theta \frac{\partial Y}{\partial \theta}\right) + \frac{1}{Y \sin^2\theta}\frac{\partial^2 Y}{\partial \phi^2}\right\} = \beta \tag{3.2.5}$$

and

$$\frac{1}{2R}\frac{d}{dr}\left(r^2\frac{dR}{dr}\right) + \left(\varepsilon - \frac{A}{r} - \frac{B}{r^2}\right) = \beta \tag{3.2.6}$$

The angular equation (3.2.5) has the same problem as for hydrogen atom. Apparently, the angular equation and its solution are unrelated to the specific form of $V(r_\mu)$. It is common to central-force field. This is very important as it makes the angular problem of the Weakest Bound Electron (such as bonding power, bonding direction, etc.) be able to be treated as the angular problem of hydrogen atom.

Table 3.1 lists $Y_{l,m}(\theta, \phi)$ functions of the Weakest Bound Electron.

By solving the angular equation, we get

$$\beta = l(l+1) \tag{3.2.7}$$

$$l = 0, 1, 2, \cdots, n-1$$

3.2 The Solution of the Radial Equation

Table 3.1 $Y_{l,m}(\theta, \phi)$ functions [1] of the Weakest Bound Electron

$Y_{0,0} = s = \sqrt{\dfrac{1}{4\pi}}$

$Y_{1,0} = p_z = \sqrt{\dfrac{3}{4\pi}} \cos\theta$

$Y_{1,\pm1} = \begin{cases} p_x = \sqrt{\dfrac{3}{4\pi}} \sin\theta \cos\phi \\ p_y = \sqrt{\dfrac{3}{4\pi}} \sin\theta \sin\phi \end{cases}$

$Y_{2,0} = d_{z^2} = \sqrt{\dfrac{5}{16\pi}} \left(3\cos^2\theta - 1\right)$

$Y_{2,\pm1} = \begin{cases} d_{xz} = \sqrt{\dfrac{15}{4\pi}} \sin\theta \cos\theta \cos\phi \\ d_{yx} = \sqrt{\dfrac{15}{4\pi}} \sin\theta \cos\theta \sin\phi \end{cases}$

$Y_{2,\pm2} = \begin{cases} d_{xy} = \sqrt{\dfrac{15}{16\pi}} \sin^2\theta \sin 2\phi \\ d_{x^2-y^2} = \sqrt{\dfrac{15}{16\pi}} \sin^2\theta \cos 2\phi \end{cases}$

$Y_{3,0} = f_{z^3} = \sqrt{\dfrac{7}{16\pi}} \left(5\cos^3\theta - 3\cos\theta\right)$

$Y_{3,\pm1} = \begin{cases} f_{xz^2} = \sqrt{\dfrac{21}{32\pi}} \sin\theta \left(5\cos^2\theta - 1\right) \cos\phi \\ f_{yz^2} = \sqrt{\dfrac{21}{32\pi}} \sin\theta \left(5\cos^2\theta - 1\right) \sin\phi \end{cases}$

$Y_{3,\pm2} = \begin{cases} f_{z(x^2-y^2)} = \sqrt{\dfrac{105}{16\pi}} \sin^2\theta \cos\theta \cos 2\phi \\ f_{zxy} = \sqrt{\dfrac{105}{16\pi}} \sin^2\theta \cos\theta \sin 2\phi \end{cases}$

$Y_{3,\pm3} = \begin{cases} f_{x(x^2-3y^2)} = \sqrt{\dfrac{35}{32\pi}} \sin^3\theta \cos 3\phi \\ f_{y(3x^2-y^2)} = \sqrt{\dfrac{35}{32\pi}} \sin^3\theta \sin 3\phi \end{cases}$

After substituting (3.2.7) into (3.2.6) followed by reorganization, we have

$$\frac{1}{2}\frac{d^2 R}{dr^2} + \frac{1}{r}\frac{dR}{dr} + \left[\varepsilon - \frac{A}{r} - \frac{B}{r^2} - \frac{l(l+1)}{2r^2}\right] R = 0 \quad (3.2.8)$$

Let

$$2B + l(l+1) = l'(l'+1) \quad (3.2.9)$$

and substitute (3.2.2) into it, then

$$\frac{1}{2}\frac{d^2R}{dr^2} + \frac{1}{r}\frac{dR}{dr} + \left[\varepsilon + \frac{Z'}{r} - \frac{l'(l'+1)}{2r^2}\right]R = 0 \quad (3.2.10)$$

3.2.2 Generalized Laguerre Functions

When the radial equation (3.2.10) is solved by generalized Laguerre functions given in reference [1], we will get the expression of the Weakest Bound Electron at its bound state for energy ε as well as for the radial wave function in a form of generalized Laguerre polynomial:

$$\varepsilon = -\frac{Z'^2}{2n'^2} \quad (3.2.11)$$

$$R(r) = A e^{-\frac{Z'r}{n'}} r^{l'} \mathbf{L}_{n-l-1}^{2l'+1}\left(\frac{2Z'r}{n'}\right) \quad (3.2.12)$$

In Eqs. (3.2.11) and (3.2.12), Z', n', l', A, and $\mathbf{L}_{n-l-1}^{2l'+1}\left(\frac{2Z'r}{n'}\right)$, are effective nuclear charge, effective principle quantum number, effective azimuthal quantum number, normalization factor and generalized Laguerre polynomial, respectively.

$$n' = k + l' + 1 \quad (3.2.13)$$

k is the number of terms remained after series were discontinued in the process of solving Eq. (3.2.10).

$$n' = k + l' + 1 = k + l + 1 + l' - l = n + l' - l \quad (3.2.14)$$

where n and l are principle quantum number and azimuthal quantum number, respectively.

From Eq. (3.2.9), we have

$$l'(l'+1) - [l(l+1) + 2B] = 0 \quad (3.2.15)$$

$$l' = -\frac{1}{2} \pm \frac{1}{2}\sqrt{(2l+1)^2 + 8B}. \quad (3.2.16)$$

3.2 The Solution of the Radial Equation

Choose the plus sign, then

$$l' = -\frac{1}{2} + \frac{1}{2}\sqrt{(2l+1)^2 + 8B}. \quad (3.2.17)$$

$$(2l'+1)^2 = (2l+1)^2 + 8B \quad (3.2.18)$$

After reorganization, we get

$$B = \frac{(l'+l+1)(l'-l)}{2} \quad (3.2.19)$$

Let

$$l' = l + d \quad (3.2.20)$$

then

$$B = \frac{d(d+1) + 2dl}{2} \quad (3.2.21)$$

Substitute (3.2.20) into (3.2.14), then

$$n' = n + d \quad (3.2.22)$$

Thus the parameter d is used to convert integral principle quantum number n and azimuthal quantum number l into non-integral effective principle quantum number n' and effective azimuthal quantum number l'.

Using normalization condition

$$\int_0^\infty [R(r)]^2 r^2 dr = 1 \quad (3.2.23)$$

can determine the normalization factor A in Eq. (3.2.12). And then

$$|A|^2 \int_0^\infty e^{-2Z'r/n'} r^{2l'+2} \left[L_{n-l-1}^{2l'+1}\left(\frac{2Z'r}{n'}\right) \right]^2 dr = 1 \quad (3.2.24)$$

or

$$|A|^2 \left(\frac{n'}{2Z'}\right)^{2l'+3} \int_0^\infty \left(\frac{2Z'r}{n'}\right)^{2l'+2} e^{-2Z'r/n'}$$
$$\times \left[L_{n-l-1}^{2l'+1}\left(\frac{2Z'r}{n'}\right)\right]^2 d\left(\frac{2Z'r}{n'}\right) = 1 \quad (3.2.25)$$

Using an integral formula containing the product of two generalized Laguerre polynomials (see next section)

$$dZ = (-1)^{n+n'} \Gamma(\lambda+1) \sum_k \binom{\lambda-\mu}{n-\kappa}\binom{\lambda-\mu'}{n'-\kappa}\binom{\lambda+k}{k} \cdot R_e(\lambda) > -1 \quad (3.2.26)$$

as long as Eqs. (3.2.24) or (3.2.25) satisfies

$$2l' + 2 > -1 \quad (3.2.27)$$

i.e.

$$l' > -\frac{3}{2} = -1.5 \quad (3.2.28)$$

the integral on the left side of (3.2.24) or (3.2.25) is integrable, then the normalization factor A can be obtained.

Table 3.2 gives a general expression for the radial wave function $R_{n',l'}(r)$ of the Weakest Bound Electron.

Substituting (3.2.21) back into (3.1.5) leads to

$$V(r_\mu) = -\frac{Z'_\mu}{r_\mu} + \frac{d_\mu(d_\mu+1) + 2d_\mu l_\mu}{2r_\mu^2} \quad (3.2.29)$$

or with subscript neglected,

$$V(r) = -\frac{Z'}{r} + \frac{d(d+1) + 2dl}{2r^2} \quad (3.2.30)$$

This is the functional form for the non-relativistic approximate analytical potential of the Weakest Bound Electron μ in a central force field after considering shielding effect, penetration effect and polarization. Its physical meaning is that the first term of above equation represents the potential energy in a central force field of point charge with effective nuclear charge $+Z'e$ which is formed by the shielding effect of $(N-\mu)$ Non-Weakest Bound Electrons from nucleus and orbital penetration effect of electrons; the second term represents the potential energy in a dipole field due to

3.2 The Solution of the Radial Equation

Table 3.2 $R_{n',l'}(r)$ has a general expression $\left(\rho = \frac{2Z'r}{n'a_0}\right)$ [1]

n	l	$R_{n',l'}(r)$
1	0	$[\Gamma(2l'+3)]^{-\frac{1}{2}}\left(\frac{2}{n'}\right)^{\frac{3}{2}}\left(\frac{Z'}{a_0}\right)^{\frac{3}{2}}\rho^{l'}e^{-\frac{\rho}{2}}$ ①
2	0	$[\Gamma(2l'+3)]^{-\frac{1}{2}}[2l'+4]^{-\frac{1}{2}}\left(\frac{2}{n'}\right)^{\frac{3}{2}}\left(\frac{Z'}{a_0}\right)^{\frac{3}{2}}\rho^{l'}[(2l'+2)-\rho]e^{-\frac{\rho}{2}}$
	1	$[\Gamma(2l'+3)]^{-\frac{1}{2}}\left(\frac{2}{n'}\right)^{\frac{3}{2}}\left(\frac{Z'}{a_0}\right)^{\frac{3}{2}}\rho^{l'}e^{-\frac{\rho}{2}}$
3	0	$[\Gamma(2l'+3)]^{-1/2}\left[\frac{1}{4}+(2l'+3)+\frac{(2l'+4)(2l'+3)}{2}\right]^{-1/2}$ $\times\left(\frac{2}{n'}\right)^{3/2}\left(\frac{Z'}{a_0}\right)^{3/2}\rho^{l'}\left[\frac{(2l'+3)(2l'+2)}{2}\right.$ $\left.-(2l'+3)\rho+\frac{1}{2}\rho^2\right]e^{-\rho/2}$
	1	$[\Gamma(2l'+3)]^{-\frac{1}{2}}[2l'+4]^{-\frac{1}{2}}\left(\frac{2}{n'}\right)^{\frac{3}{2}}\left(\frac{Z'}{a_0}\right)^{\frac{3}{2}}\rho^{l'}[(2l'+2)-\rho]e^{-\frac{\rho}{2}}$
	2	$[\Gamma(2l'+3)]^{-\frac{1}{2}}\left(\frac{2}{n'}\right)^{\frac{3}{2}}\left(\frac{Z'}{a_0}\right)^{\frac{3}{2}}\rho^{l'}e^{-\frac{\rho}{2}}$
4	0	$[\Gamma(2l'+3)]^{-1/2}\left[\frac{1}{36}+\frac{(2l'+3)}{4}+\frac{(2l'+4)(2l'+3)}{2}\right.$ $\left.+\frac{(2l'+5)(2l'+4)(2l'+3)}{6}\right]^{-1/2}\left(\frac{2}{n'}\right)^{3/2}$ $\times\left(\frac{Z'}{a_0}\right)^{3/2}\rho^{l'}\left[\frac{(2l'+4)(2l'+3)(2l'+2)}{6}\right.$ $\left.-\frac{(2l'+4)(2l'+3)}{2}\rho+\frac{(2l'+4)}{2}\rho^2-\frac{1}{6}\rho^3\right]e^{-\rho/2}$
	1	$[\Gamma(2l'+3)]^{-1/2}\left[\frac{1}{4}+(2l'+3)+\frac{(2l'+4)(2l'+3)}{2}\right]^{-1/2}$ $\times\left(\frac{2}{n'}\right)^{3/2}\left(\frac{Z'}{a_0}\right)^{3/2}\rho^{l'}\left[\frac{(2l'+3)(2l'+2)}{2}\right.$ $\left.-(2l'+3)\rho+\frac{1}{2}\rho^2\right]e^{-\rho/2}$
	2	$[\Gamma(2l'+3)]^{-\frac{1}{2}}[2l'+4]^{-\frac{1}{2}}\left(\frac{2}{n'}\right)^{\frac{3}{2}}\left(\frac{Z'}{a_0}\right)^{\frac{3}{2}}\rho^{l'}[(2l'+2)-\rho]e^{-\frac{\rho}{2}}$

(continued)

the polarization of "core$_\mu$" consisting of Non-Weakest Bound Electrons and nucleus by the Weakest Bound Electron μ.

Substituting (3.2.30) back into (3.1.8) leads to

$$\left[-\frac{1}{2}\nabla^2-\frac{Z'}{r}+\frac{d(d+1)+2dl}{2r^2}\right]\varphi=\varepsilon\varphi \quad (3.2.31)$$

This is the analytical expression of one-electron Schrödinger equation of the Weakest Bound Electron μ under above approximations.

Table 3.2 (continued)

n	l	$R_{n',l'}(r)$
	3	$[\Gamma(2l'+3)]^{-\frac{1}{2}}\left(\frac{2}{n'}\right)^{\frac{3}{2}}\left(\frac{Z'}{a_0}\right)^{\frac{3}{2}}\rho^{l'}e^{-\frac{\rho}{2}}$
5	0	$[\Gamma(2l'+3)]^{-1/2}\left[\dfrac{1}{2304}+\dfrac{(2l'+3)}{36}+\dfrac{(2l'+4)(2l'+3)}{8}\right.$ $+\dfrac{(2l'+5)(2l'+4)(2l'+3)}{6}$ $\left.+\dfrac{(2l'+6)(2l'+5)(2l'+4)(2l'+3)}{24}\right]^{-1/2}$ $\times\left(\dfrac{2}{n'}\right)^{3/2}\left(\dfrac{Z'}{a_0}\right)^{3/2}\rho^{l'}$ $\times\left[\dfrac{(2l'+5)(2l'+4)(2l'+3)(2l'+2)}{24}\right.$ $-\dfrac{(2l'+5)(2l'+4)(2l'+3)}{6}\rho+\dfrac{(2l'+5)(2l'+4)}{4}\rho^2$ $\left.-\dfrac{(2l'+5)}{6}\rho^3+\dfrac{1}{24}\rho^4\right]e^{-\rho/2}$
	1	$[\Gamma(2l'+3)]^{-1/2}\left[\dfrac{1}{36}+\dfrac{(2l'+3)}{4}+\dfrac{(2l'+4)(2l'+3)}{2}\right.$ $\left.+\dfrac{(2l'+5)(2l'+4)(2l'+3)}{6}\right]^{-1/2}\left(\dfrac{2}{n'}\right)^{3/2}$ $\times\left(\dfrac{Z'}{a_0}\right)^{3/2}\rho^{l'}\left[\dfrac{(2l'+4)(2l'+3)(2l'+2)}{6}\right.$ $\left.-\dfrac{(2l'+4)(2l'+3)}{2}\rho+\dfrac{(2l'+4)}{2}\rho^2-\dfrac{1}{6}\rho^3\right]e^{-\rho/2}$
	2	$[\Gamma(2l'+3)]^{-1/2}\left[\dfrac{1}{4}+(2l'+3)+\dfrac{(2l'+4)(2l'+3)}{2}\right]^{-1/2}$ $\times\left(\dfrac{2}{n'}\right)^{3/2}\left(\dfrac{Z'}{a_0}\right)^{3/2}\rho^{l'}\left[\dfrac{(2l'+3)(2l'+2)}{2}\right.$ $\left.-(2l'+3)\rho+\dfrac{1}{2}\rho^2\right]e^{-\rho/2}$
	3	$[\Gamma(2l'+3)]^{-\frac{1}{2}}[2l'+4]^{-\frac{1}{2}}\left(\dfrac{2}{n'}\right)^{\frac{3}{2}}\left(\dfrac{Z'}{a_0}\right)^{\frac{3}{2}}\rho^{l'}[(2l'+2)-\rho]e^{-\frac{\rho}{2}}$

Note: ① is applicable to a many-electron atomic or ionic system

3.2 The Solution of the Radial Equation

By solving Eq. (3.2.31), the obtained expressions for one-electron energy ε and the radial wave function of the Weakest Bound Electron μ are

$$-\frac{Z'^2}{2n'^2} \quad (3.2.32)$$

$$R(r) = A e^{-Z'r/n'} r^{l'} \mathbf{L}_{n-l-1}^{2l'+1}\left(\frac{2Z'r}{n'}\right) \quad (3.2.33)$$

ε_μ^0 is the result under the condition of non-relativity and central-force field, so there should be

$$-I_\mu = \varepsilon_\mu = \varepsilon_\mu^0 + \Delta E_c + \Delta E_r \quad (3.2.34)$$

where ΔE_r and ΔE_c represent the contributions to the one-electron energy of the Weakest Bound Electron μ from neglected relativity and electron correlation effects, respectively. Since these corrections are small, we have

$$\varepsilon_\mu^0 \approx -I_\mu \quad (3.2.35)$$

Again because ε represents non-relativistic energy of the Weakest Bound Electron in a central force field with given analytical potential, it is different from ε_μ^0, but the difference is very small, so $\varepsilon \approx \varepsilon_\mu^0$ and $\varepsilon \approx -I_\mu$.

In our previous studies, the result that $V(r)$ is in the form of (3.2.30) and one-electron Schrodinger equation is in the form of (3.2.31) was called Weakest Bound Electron Potential Model Theory (WBEPM Theory). The name is still used in this book. Previous discussion shows that WBEPM Theory can only be applied to atoms. It is one expression of WBE Theory in an atomic system with given analytical potential.

Both Luya Wang, et al. and Zidong Chen, et al. gave recurrence relation of the radial wave function based on WBEPM Theory, which makes calculations more convenient.

3.2.3 Restore the Form of Hydrogen and Hydrogen-Like Atoms

In Quantum mechanics, hydrogen and hydrogen-like atoms are one-electron systems which can be solved rigorously. The only electron in the system is the Weakest Bound Electron of the system. Since there is no interaction between electrons in hydrogen and hydrogen-like atoms, $Z' = Z, d = 0$. Then

$$n' = n + d = n \quad (3.2.36)$$

$$l' = l + d = l \tag{3.2.37}$$

This gives rise to

$$-\frac{Z^2}{2n^2} \tag{3.2.38}$$

and

$$u\frac{d^2M}{du^2} + (k+1-u)\frac{dM}{du} + (n-l-1)M = 0 \tag{3.2.39}$$

or

$$u\frac{d^2M}{du^2} + (k+1-u)\frac{dM}{du} + (n+l-k)M = 0 \tag{3.2.40}$$

In above equation, k is an integer ($k = 2l + 1$), while (3.2.39) or (3.2.40) is a differential equation that readers are familiar with and that the associated Laguerre polynomials must satisfy when solving equation $R(r)$ for hydrogen and hydrogen-like atoms in quantum mechanics. Its normalized solution is

$$\begin{aligned}R_{nl}(r) = -&\left\{\frac{(n-l-1)!}{2n[(n+l)!]^3}\right\}^{\frac{1}{2}}\left(\frac{2Z}{n}\right)^{\frac{3}{2}} \\ \times &\left(\frac{2Zr}{n}\right)^l e^{-Zr/n}\mathbf{L}_{n+l}^{2l+1}\left(\frac{2Zr}{n}\right)\end{aligned} \tag{3.2.41}$$

WBEPM Theory can naturally restore the results of quantum mechanics for hydrogen and hydrogen-like atoms. This indicates that WBEPM Theory is a unified theoretical model that can be generally applied to one-electron and many-electron systems. Generalized Laguerre polynomial for solving the radial wave function is a more general method. Associated Laguerre polynomial for solving one-electron atomic and ionic system is a special case of generalized Laguerre polynomial.

3.2.4 The Definition and Properties of Generalized Laguerre Functions

3.2.4.1 Γ(Gamma) Function [1, 12, 13]

Since WBEPM Theory involves knowledge related to Γ function, here we first describe the definition and common properties of Γ function.

3.2 The Solution of the Radial Equation

1. **The definition of Γ function**

   Γ function defined by integral can be expressed as

   $$\Gamma(Z) = \int_0^\infty e^{-t} t^{z-1} dt \qquad (3.2.42)$$

   This definition can only be applied to the region of $R_e(Z) > 0$ since this is the condition under which the integral converges at $t = 0$. The integral on the right side of Eq. (3.2.24) is called Euler integral of the second kind.

   A more general definition of gamma function $\Gamma(Z)$ is

   $$\begin{aligned}\Gamma(Z) &= \lim_{n\to\infty} \frac{1\times 2\cdots(n-1)}{Z(Z+1)\cdots(Z+n-1)} n^z \\ &= \frac{1}{Z}\prod_{n=1}^\infty \left[\left(1+\frac{Z}{n}\right)^{-1}\left(1+\frac{1}{n}\right)^z\right]\end{aligned} \qquad (3.2.43)$$

   Except for a pole $Z = -n$, the above formula is tenable for any value of Z.

2. **Recurrence relation**

   $\Gamma(Z)$ satisfies the following recurrence relation

   $$\Gamma(Z+1) = Z\Gamma(Z) \qquad (3.2.44)$$

   The recurrence relation can be proved as follows. According to the integral definition of gamma function, there is

   $$\Gamma(Z+1) = \int_0^\infty e^{-t} t^z dt \qquad (3.2.45)$$

   Using integration by parts, the integral on the right side of above equation becomes

   $$\int_0^\infty e^{-t} t^z dt = [-e^{-t} t^z]_0^\infty + z\int_0^\infty e^{-t} t^{z-1} dt = Z\Gamma(Z) \qquad (3.2.46)$$

Assume n is a random positive integer, and Eq. (3.2.44) can be further extended to

$$\Gamma(Z+n) = (Z+n-1)(Z+n-2)\cdots(Z+1)Z\Gamma(Z) \qquad (3.2.47)$$

or

$$\Gamma(Z) = \frac{\Gamma(Z+n)}{z(z+1)\cdots(Z+n-1)} = \frac{1}{(Z)_n}\int_0^\infty e^{-t}t^{z+n-1}dt \qquad (3.2.48)$$

where

$$(Z)_n = Z(Z+1)\cdots(Z+n-1) \qquad (3.2.49)$$

Equation (3.2.48) extends the definition of $\Gamma(Z)$ to the region of $R_e(Z) > -n$. From Eq. (3.2.48), we can see that $\Gamma(Z)$ is a meromorphic function of Z with poles at $Z = -n (n = 0, 1, 2, \cdots)$. The residues of the function at these poles are:

$$\lim_{z \to -n}(Z+n)\Gamma(Z) = \frac{\Gamma(Z+n+1)}{z(z+1)\cdots(Z+n-1)}\bigg|_{z=-n} = \frac{(-1)^n}{n!} \qquad (3.2.50)$$

3. **Related formulas of Γ function**

$$\Gamma(n+1) = n! \quad (n \text{ is a positive integer}) \qquad (3.2.51)$$

Specifically,

$$\Gamma(1) = 1 \qquad (3.2.52)$$

$$\Gamma(2) = 1 \qquad (3.2.53)$$

$$\Gamma(Z)\Gamma(-Z) = -\frac{\pi}{Z \sin \pi Z} \qquad (3.2.54)$$

$$\Gamma(n+Z)\Gamma(n-Z) = \frac{\pi Z}{\sin \pi Z}[(n-1)!]^2 \prod_{k=1}^{n-1}\left(1 - \frac{Z^2}{k^2}\right) (n = 1, 2, 3, \cdots) \qquad (3.2.55)$$

$$\Gamma(Z)\Gamma(1-Z) = \frac{\pi}{\sin \pi Z} \qquad (3.2.56)$$

3.2 The Solution of the Radial Equation

In Eq. (3.2.56), if let $Z = 1/2$, there is

$$\Gamma\left(\frac{1}{2}\right) = \sqrt{\pi} \qquad (3.2.57)$$

4. **When $1 \le Z \le 2$, the approximate value of $\Gamma(Z)$**

   Usually, this value is collected in and can be referred to <<Mathematical Handbook>>

5. **Incomplete gamma functions [13]**

   The definition and related expression of incomplete gamma functions are shown below.

$$\gamma(v, Z) = \int_0^Z u^{v-1} e^{-u} du \quad [|Z|\langle\infty, R_e(v)\rangle 0] \qquad (3.2.58)$$

$$\Gamma(v, Z) = \Gamma(v) - \gamma(v, Z) = \int_Z^\infty u^{v-1} e^{-u} du \quad (|\arg Z| < \pi) \qquad (3.2.59)$$

$$\gamma(\alpha + 1, Z) = \alpha\gamma(\alpha, Z) - Z^\alpha e^{-Z} \qquad (3.2.60)$$

$$\Gamma(\alpha + 1, Z) = \alpha\Gamma(\alpha, Z) + Z^\alpha e^{-Z} \qquad (3.2.61)$$

$$\Gamma(\alpha, Z) = \frac{e^{-Z} Z^\alpha}{\Gamma(1-\alpha)} \int_0^\infty \frac{e^{-t} t^{-a}}{Z+t} dt \qquad (3.2.62)$$

3.2.4.2 Generalized Laguerre Polynomial and Its Related Integral [1, 12, 13]

1. **Laguerre polynomial**

   Laguerre polynomial $L_\alpha(x)$ is a α-th degree polynomial of x. It can be defined by the following formula:

$$L_\alpha(x) = e^x \frac{d^\alpha}{dx^\alpha}\left(x^\alpha e^{-x}\right) \qquad (3.2.63)$$

Laguerre polynomial satisfies Laguerre differentiation equation

$$x\frac{d^2 L_\alpha}{dx^2} + (1-x)\frac{dL_\alpha}{dx} + \alpha L_\alpha = 0 \qquad (3.2.64)$$

2. Associated Laguerre polynomial

The β-th derivative of $L_\alpha(x)$ is called associated Laguerre polynomial and denoted by $L_\alpha^\beta(x)$.

$$L_\alpha^\beta(x) = \frac{d^\beta}{dx^\beta} L_\alpha(x) = \frac{d^\beta}{dx^\beta}\left[e^x \frac{d^\alpha}{dx^\alpha}(x^\alpha e^{-x})\right] \quad (\beta \leq \alpha) \tag{3.2.65}$$

It is a polynomial of (α-β) degree, in which β is a positive integer or zero and is less than or equal to α.

The differential equation that $L_\alpha^\beta(x)$ satisfies is

$$x\frac{d^2}{dx^2}L_\alpha^\beta + (\beta + 1 - x)\frac{d}{dx}L_\alpha^\beta + (\alpha - \beta)L_\alpha^\beta = 0 \tag{3.2.66}$$

3. Generalized Laguerre polynomial

(1) The definition of generalized laguerre polynomial is

$$L_n^\mu(Z) = \frac{\Gamma(\mu + 1 + n)}{n!\Gamma(\mu + 1)} F(-n, \mu + 1, Z) \tag{3.2.67}$$

This is a polynomial of n degree, in which μ is any real number or complex number that is not a negative integer; $F(-n, \mu+1, Z)$ is Kummer's function.

If $\mu = 0$, $L_n^0(Z) = L_n(Z)$. This is Laguerre polynomial that we described before.

If $\mu = m$ (m is a positive integer), $L_n^\mu(Z) = L_n^m(Z)$. This is associated Laguerre polynomial that we described before.

(2) The expressions for the integral and derivative of generalized Laguerre polynomial are

$$\begin{aligned}L_n^\mu(Z) &= \frac{e^z Z^{-\mu}}{n!}\frac{d^n}{dZ^n}(Z^{\mu+n}e^{-Z}) \\ &= \sum_{k=0}^n (-1)^k \binom{n+\mu}{n-k}\frac{Z^k}{k!} \\ &= \frac{(-1)^n}{2\pi i}\int^{(0+)} e^{Zt}(1-t)^{\mu+n}t^{-n-1}dt\end{aligned} \tag{3.2.68}$$

(3) Generalized Laguerre differential equation

Generalized Laquerre polynomial satisfies the following differential equation:

$$ZL_n^{\mu''}(Z) + (\mu + 1 - Z)L_n^{\mu'}(Z) + nL_n^\mu(Z) = 0 \tag{3.2.69}$$

This equation is called generalized Laguerre differential equation.

3.2 The Solution of the Radial Equation

(4) Generating function

$$\frac{e^{-Zt/(1-t)}}{(1-t)^{\mu+1}} = \sum_{n=0}^{\infty} L_n^{\mu}(Z) t^n \quad (|t| < 1) \tag{3.2.70}$$

(5) The integral of the product of two generalized Laguerre polynomials

$$\int_0^{\infty} Z^{\lambda} e^{-Z} L_n^{\mu}(Z) L_{n'}^{\mu'}(Z) dZ = (-1)^{n+n'} \Gamma(\lambda+1) \sum_k \binom{\lambda-\mu}{n-k} \binom{\lambda-\mu'}{n'-k} \binom{\lambda+k}{k} \tag{3.2.71}$$

where $R_e(\lambda) > -1$ to ensure that the integral converges at the lower limit.

This is an important formula of integral of product which can be proved in this way [12]:

From (3.2.71)

$$\sum_{n=0}^{\infty} t^n L_n^{\mu}(Z) \sum_{n'=0}^{\infty} s^{n'} L_{n'}^{\mu'}(Z)$$

$$= \sum_{n,n'} t^n s^{n'} L_n^{\mu}(Z) L_{n'}^{\mu'}(Z) = \frac{e^{-Z\left(\frac{t}{1-t} - \frac{s}{1-s}\right)}}{(1-t)^{\mu+1}(1-s)^{\mu'+1}} \tag{3.2.72}$$

Assume $|s| < 1$, $|t| < 1$, then

$$\sum_{n,n'} t^n s^{n'} \int_0^{\infty} Z^{\lambda} e^{-Z} L_n^{\mu}(Z) L_{n'}^{\mu'}(Z) dZ$$

$$= \int_0^{\infty} Z^{\lambda} \frac{e^{-\frac{Z(1-ts)}{(1-t)(1-s)}}}{(1-t)^{\mu+1}(1-s)^{\mu'+1}} dZ$$

$$= (1-t)^{\lambda-\mu}(1-s)^{\lambda-\mu'}(1-ts)^{-\lambda-1} \int_0^{\infty} e^{-v} v^{\lambda} dv$$

$$= \Gamma(\lambda+1) \sum_l \binom{\lambda-\mu}{l}(-t)^l \sum_{l'} \binom{\lambda-\mu'}{l'}(-s)^{l'}$$

$$\times \sum_k \binom{-\lambda-1}{k}(-ts)^k$$

$$= \Gamma(\lambda+1) \sum_{n,n'} t^n s^{n'} \sum_k \binom{\lambda-\mu}{n-k}\binom{\lambda-\mu'}{n'-k}\binom{-\lambda-1}{k}(-1)^{n+n'+k} \tag{3.2.73}$$

After comparing the coefficient of $t^n s^{n'}$ on both sides of the equation and noting that

$$(-1)^k \binom{-\lambda - 1}{k} = \binom{\lambda + k}{k}$$

We know (3.2.71) has been proved.

(6) Expansion formula [12]

$$Z^s L_n^\mu(Z) = \sum_{r=0}^{n+s} \alpha_r^s L_{n+s-r}^{\mu+p}(Z) \tag{3.2.74}$$

Expansion coefficient

$$\alpha_r^s = (-1)^{s+r} \frac{(n+s-r)!\Gamma(s+\mu+p+1)}{\Gamma(n+s+\mu+p-r+1)}$$
$$\times \sum_k \binom{s+p}{n-k} \binom{s}{k+r-n} \binom{s+\mu+p+k}{k} \tag{3.2.75}$$

(7) Orthogonality of generalized Laguerre polynomier

$$\int_0^\infty \frac{x^\alpha}{e^x} L_m^\alpha(x) L_n^\alpha(x) dx = \begin{cases} 0 & m \neq n, a > -1 \\ \frac{\Gamma(n+\alpha+1)}{n!} & m = n, a > -1 \end{cases}$$
$$\left(\text{or} \quad = \frac{\Gamma(n+\alpha+1)}{n!} \delta_{mn} \right) \tag{3.2.76}$$

3.2.5 The Proof of the Satisfaction of Hellmann–Feynman Theorem

The content of generalized differential Hellmann–Feynman theorem is that for a quantum mechanics system, if ψ is a normalized correct wave function of Hamiltonian $\widehat{H}$ and E is the corresponding eigenvalue, the first derivative of E with respect to any parameter involved in $\widehat{H}$ is equal to the expectation value of the first derivative of $\widehat{H}$ with respective to that parameter. The parameter can be the one included in approximations, or physical quantity of the system such as nuclear charge.

Below we will prove whether Hellmann–Feynman theorem is satisfied when the potential function of the Weakest Bound Electron is in the analytical form described before.

For the Weakest Bound Electron μ, in the central field approximation, its potential function is

$$V(r_\mu) = -\frac{Z'_\mu}{r_\mu} + \frac{d_\mu(d_\mu + 1) + 2d_\mu l_\mu}{2r_\mu^2} \tag{3.2.77}$$

3.2 The Solution of the Radial Equation

and its corresponding one-electron Schrodinger equation is

$$\hat{H}''_\mu \psi''_i(\mu) = \left(-\frac{1}{2}\nabla^2_\mu - \frac{Z'_\mu}{r_\mu} + \frac{d_\mu(d_\mu+1) + 2d_\mu l_\mu}{2r^2_\mu}\right)\psi''_i(\mu) = \varepsilon''_\mu \psi''_i(\mu) \quad (3.2.78)$$

where $\hat{\mu}''$ represents the non-relativistic one-electron Hamiltonian operator of the Weakest Bound Electron μ with a given analytical potential; $\psi''_i(\mu)$ and ε''_μ are the eigenfunction and eigenvalue of the operator, respectively.

After neglecting the superscript and subscript of Eq. (3.2.78) and solving it rigorously, people can get an important set of expressions in the following, including expressions for eigen energy ε, radial wave function $R_{n',l'}$, matrix element and expectation value of radial position operator r_k^μ, etc. Among these expressions, those related to the proof here are

$$\varepsilon = -\frac{Z'^2}{2n'^2} \quad (3.2.79)$$

$$\frac{1}{r_\mu} = \frac{Z'}{n'^2} \quad (3.2.80)$$

$$\frac{1}{r^2} = \frac{2Z'}{n'^3(2l'+1)} \quad (3.2.81)$$

First, considering Z' as a parameter, we have

$$\frac{\partial H}{\partial Z'} = -\frac{1}{r} \quad (3.2.82)$$

$$\frac{\partial H}{\partial Z'} = -\frac{1}{r} = -\frac{Z'}{n'^2} \quad (3.2.83)$$

and

$$\frac{\partial \varepsilon}{\partial Z'} = -\frac{Z'}{n'^2} \quad (3.2.84)$$

Comparing (3.2.83) and (3.2.84), we have

$$\frac{\partial H}{\partial Z'} = \frac{\partial \varepsilon}{\partial Z'} \quad (3.2.85)$$

Again consider d_μ as the parameter, then

$$\frac{2d+1+2l}{2r^2} \quad (3.2.86)$$

$$\frac{\partial H}{\partial d} = \frac{2d+1+2l}{2}\frac{1}{r^2} = \frac{Z'^2}{n'^3} \qquad (3.2.87)$$

and

$$\frac{\partial \varepsilon}{\partial d} = \frac{\partial}{\partial d}\left(-\frac{Z'^2}{2n'^2}\right) = \frac{\partial}{\partial d}\left[-\frac{Z'^2}{2(n+d)^2}\right] = \frac{Z'^2}{(n+d)^3} = \frac{Z'^2}{n'^3} \qquad (3.2.88)$$

Comparing (3.2.87) and (3.2.88), we get

$$\frac{\partial H}{\partial d} = \frac{\partial \varepsilon}{\partial d} \qquad (3.2.89)$$

From (3.2.85) and (3.2.89), it is known that Hellmann–Feynman theorem is satisfied.

3.3 Matrix Element and Mean Value of Radial Operator r^k

According to Taylor expansion and differential property of generalized Laguerre polynomial, Wen et al. [14] derived general formulae to calculate matrix elements and average value of a random power of the radial position operator r_k^μ within WBEPM Theory. The results are

$$\langle n_1 l_1 | r^k | n_2 l_2 \rangle$$

$$= (-1)^{n_1+n_2+l_1+l_2}\left(\frac{2Z_1'}{n_1'}\right)^{l_1'}\left(\frac{2Z_2'}{n_2'}\right)^{l_2'}$$

$$\times \left(\frac{Z_1'}{n_1'}+\frac{Z_2'}{n_2'}\right)^{-l_1'-l_2'-k-3}\left[\frac{n_1'^4 \Gamma(n_1'+l_1'+1)}{4Z_1'^3(n_1-l_1-1)!}\right]^{-\frac{1}{2}}$$

$$\times \left[\frac{n_2'^4 \Gamma(n_2'+l_2'+1)}{4Z_2'^3(n_2-l_2-1)!}\right]^{-\frac{1}{2}} \sum_{m_1=0}^{n_1-l_1-1}\sum_{m_2=0}^{n_2-l_2-1}\frac{(-1)^{m_2}}{m_1!m_2!}$$

$$\times \left(\frac{Z_1'}{n_1'}-\frac{Z_2'}{n_2'}\right)^{m_1+m_2}\left(\frac{Z_1'}{n_1'}+\frac{Z_2'}{n_2'}\right)^{-m_1-m_2}$$

$$\times \Gamma(l_1'+l_2'+m_1+m_2+k+3)$$

$$\times \sum_{m_3=0}^{s}\binom{l_2'-l_1'+k+m_2+1}{n_1-l_1-1-m_1-m_3}$$

$$\times \binom{l_1'-l_2'+k+m_1+1}{n_2-l_2-1-m_2-m_3}\times \binom{l_1'+l_2'+k+2+m_1+m_2+m_3}{m_3} \qquad (3.3.1)$$

3.3 Matrix Element and Mean Value of Radial Operator r^k

where $s = \min\{n_1 - l_1 - 1 - m_1, n_1 - l_1 - 1 - m_1\}$, $k > -l'_1 - l'_2$; and

$$\langle nl|r^k|nl\rangle = (\frac{n'}{2Z'})^k \frac{(n-l-1)!\,\Gamma(2l'+k+3)}{2n'\,\Gamma(n'+l'+1)}$$
$$\times \sum_{m=s'}^{n-l-1} \binom{k+1}{n-l-1-m}\binom{2l'+k+2+m}{m}^2 \quad (3.3.2)$$

where $s' = \max\{0, n-l-1-k-1\}$.

For hydrogen and hydrogen-like atom, $Z' = Z, d = 0$ (or $n' = n, l' = l$), then above two equations can naturally transit to the forms for the situations of hydrogen and hydrogen-like atom.

Zhou also derived a general formula to calculate matrix elements of a random power of the radial position operator r_k^μ within WBEPM Theory using two equivalent expressions of generalized Laguerre polynomial and integration by parts. The result is the same as Eq. (3.3.1) except that one summation is not calculated. The general formula to calculate the average value is also the same as Eq. (3.3.2). This paper also gives specific expressions for $k = 1, -1, -2, -3$ and -4. For reader's convenience, these expressions are listed below.

$$\langle nl|r|nl\rangle = \frac{3n'^2 - l'(l'+1)}{2Z'} \quad (3.3.3)$$

$$\langle nl|r^{-1}|nl\rangle = \frac{Z'}{n'^2} \quad (3.3.4)$$

$$\langle nl|r^{-2}|nl\rangle = \frac{2Z'^2}{n'^3(2l'+1)} \quad (3.3.5)$$

$$\langle nl|r^{-3}|nl\rangle = \frac{2Z'^3}{n'^3(2l'+1)(l'+1)l'} \quad (3.3.6)$$

$$\langle nl|r^{-4}|nl\rangle = \frac{4Z'^4[3n'^2 - l'(l'+1)]}{n'^5 l'(l'+1)(2l'-1)(2l'+1)(2l'+3)} \quad (l' > -1/2) \quad (3.3.7)$$

In above formulas, Jia derived a general formula to calculate matrix elements of a random power of the radial position operator r_k^μ within WBEPM Theory by direct integration of expanded generalize Laguerre polynomial. Its expression doesn't have one summation which is included in Eq. (3.3.1). The related expression is

$$\langle n_i l_i | r | n_f l_f \rangle = \left[\frac{(n'_i - l'_i - 1)!\,(n'_f - l'_f - 1)!}{4n'_i n'_f (n'_i + l'_i)!\,(n'_f - l'_f)!} \right]^{\frac{1}{2}}$$

$$\times \left(\frac{Z'_i}{n'_i} + \frac{Z'_f}{n'_f}\right)^{-k} \sum_{m_1=0}^{n'_i-l'_i-1} \sum_{m_2=0}^{n'_f-l'_f-1} \frac{(-1)^{m_1+m_2}}{m_1! m_2!}$$

$$\times \begin{bmatrix} n'_i + l'_i \\ n'_i - l'_i - 1 - m_1 \end{bmatrix} \begin{bmatrix} n'_f + l'_f \\ n'_f - l'_f - 1 - m_2 \end{bmatrix}$$

$$\times \left[\frac{2Z'_i n'_f}{Z'_i n'_f + Z'_f n'_i}\right]^{\frac{3}{2}+l'_i+m_1} \left[\frac{2Z'_f n'_i}{Z'_i n'_f + Z'_f n'_i}\right]^{\frac{3}{2}+l'_f+m_2}$$

$$\times \Gamma(l'_i + l'_f + k + 3 + m_1 + m_2) \tag{3.3.8}$$

Wen et al. [17] also derived a recurrence relation between matrix elements of every power of the radial position operator by using Hyper-Virial theorem.

Zhou derived general formulae to calculate matrix elements $\langle n_1 l_1 | r^s \frac{d^t}{dr^t} r^{s'} | n_2 l_2 \rangle$ and average value $\langle nl | r^s \frac{d^t}{dr^t} r^{s'} | nl \rangle$ which involve radial part of differential operator $r^s \frac{d^t}{dr^t} r^{s'}$.

If we let $t = 0$, $s' = 0$, these general formulae will degenerate to Eqs. (3.3.1) and (3.3.2).

G. Chen derived a recurrence relation of the matrix elements of any operator $f(r)$ within WBEPM theory by using raising and lowering operators as well as Hyper-Virial theorem. $f(r)$ can be any power of the radial position operator r_k^μ, or exponential operator e^{sr}, or any order differential operator $\frac{d^t}{dr^t}$, or their combinations.

3.4 The Exact Solutions of Scattering States in WBEPM Theory

The theoretical studies of bound states are mainly interested in discrete energy eigenvalue, eigenstate and quantum transitions between energy states. Through discussions in Chap. 2 and discussions before this section of this chapter, we have obtained one-electron Schrodinger equation of the Weakest Bound Electron at given analytical potential. By solving this equation rigorously, the expression for discrete energy spectrum and state function can be obtained. In the following sections of this chapter, we will discuss energy spectrum, quantum transition between energy states, etc., in depth.

From discussions of hydrogen atom, people know there are continuous states besides bound states. Scattering state is one type of unbounded state which involves continuous part of the energy spectrum of the system. Using the same method for bound states, we can obtain the energy spectrum for $E > 0$ although it is continuous [20–22]. Now, we need to ask whether there are exact solutions for a scattering state within WBEPM Theory? In addition, only a complete set of functions can be used for the discussions of problems of atoms and molecules as a linear expansion of

3.4 The Exact Solutions of Scattering States in WBEPM Theory

basis functions. The bound state wave function of hydrogen atom itself can't form a complete set, however, its continuous spectrum eigenfunctions and bound state eigenfunctions together can form a complete set [23]. Then can eigenfunctions of the Weakest Bound Electron form a complete set? Chen et al. [24] presented an exact solution which was normalized on the "$k/2\pi$ scale" for the radial wave functions of scattering states and the calculation formula of phase shift within WBEPM Theory. The expression for the exact solution is

$$R_{kl} = \frac{|\Gamma(l'+1-iZ'/k)|e^{\pi Z'/2k}(2kr)^{l'+1}}{\Gamma(2l'+2)}$$
$$\times \frac{e^{-ikr}}{r} F\left(l'+1+\frac{iZ'}{k}, 2l'+2, 2ikr\right) \quad (3.4.1)$$

This formula satisfies the normalization condition

$$\int_0^\infty r^2 R_{kl}^*(r) R_{kl}(r) dr = 2\pi \delta(k'-k) \quad (3.4.2)$$

They also derived analytical formulae for calculating bound-continuous transition matrix elements $\langle n_1 l_1 | r^s | k l_2 \rangle$ and analytical formulae for calculating bound-free nonrelativistic transition matrix elements [24, 25]. These formulae are useful for dealing with scattering problems such as photoionization cross-sections, etc. Chen et al. [24] wrote in the conclusion section of the paper:"In this paper, we extended WBEPM Theory to scattering states and studied scattering properties in great detail. The normalized continuous wave function precisely on the '$k/2\pi$ scale', the calculation formula of phase shift and analytical solutions of a distorted Coulomb wave were presented. We discussed analytical properties of scattering amplitude, pointed out that the poles of scattering amplitude on complex k-plane just correspond to bound state energy levels, and derived bound state wave functions from continuous wave functions. General analytical formulae for calculating bound-continuous transition matrix elements were given." "In quantum theories, the cases that Schrodinger equation can be precisely solved are very rare. Every case with exact solutions has very important theoretical meanings and has wide applications for real problems. This work and other studies demonstrate that the bound states and scattering states in WBEPM Theory proposed by Zheng can be solved precisely. This is another example that is precisely soluble in three-dimensional space, and is universal with hydrogen-like system included as a special case. For that matter, WBEPM Theory is very satisfying, not to mention its wide applications for practical problems."

3.5 The Formula for the Calculation of Fine Structure

In subsection 2.3.3, the Hamiltonian H'_μ of the Weakest Bound Electron μ was given with Breit interaction ignored and central field approximation.
Rewrite Eq. (2.3.22)

$$H'_\mu = \frac{1}{2}[P(\mu)]^2 + V(r_\mu) - \frac{1}{8c^2}[P(\mu)]^4$$
$$+ \frac{1}{2c^2 r_\mu} \frac{dV(r_\mu)}{dr_\mu}(s_\mu l_\mu) + \frac{1}{8c^2}\nabla^2 V(r_\mu) \quad (3.5.1)$$

Then the non-relativist approximation (υ/c ≤ 1) of one-electron Dirac equation of the Weakest Bound Electron μ is

$$\{\frac{1}{2}[P(\mu)]^2 + V(r_\mu) - \frac{1}{8c^2}[P(\mu)]^4$$
$$+ \frac{1}{2c^2 r_\mu} \frac{dV(r_\mu)}{dr_\mu}(s_\mu l_\mu) + \frac{1}{8c^2}\nabla^2 V(r_\mu)\}\psi'_\mu = \varepsilon'_\mu \psi'_\mu \quad (3.5.2)$$

In the above equation, the first two terms in the brace are Hamiltonian of non-relativistic Schrodinger equation in central field approximation, designated as $\widehat{H}^0_\mu$. In WBEPM Theory, if $\widehat{H}^0_\mu$ takes formula in Eq. (3.2.31), i.e.

$$\widehat{H}''_\mu = -\frac{1}{2}\nabla^2 - \frac{Z'}{r} + \frac{d(d+1) + 2dl}{2r^2} \quad (3.5.3)$$

the eigenfunction of $\widehat{H}''_\mu$ can be solved rigorously.

The third term in the brace on the left side of Eq. (3.5.2) is mass-velocity term (a correction to the theory of relativity), denoted by $\Delta\widehat{H}_m$; the fourth term is spin-orbital coupling term, denoted by $\Delta\widehat{H}_{ls}$; the fifth term is Darwin term denoted by $\Delta\widehat{H}_d$. The last three terms are very small and can be treated as perturbations in the calculations of corresponding energy corrections ΔE_m, ΔE_{ls}, ΔE_d on the condition that the eigenfunction of $\widehat{H}''_\mu$ can be solved rigorously. Their summation is the fine structure of the energy levels of the Weakest Bound Electron μ. Chen et al. [26] derived formulae for each correction term under coupling representation. The formulae that they derived are as follows:

$$\Delta E_m = -\frac{Z'^4}{2c^2 n'^4}[\frac{1}{4} + \frac{2n'^2 + 2B - 2n'l' - n'}{n'(2l'+1)}$$
$$+ \frac{48B^2 n'^2 - 16B^2 l'(l'+1) - 16Bn'^2(2l'+3)(2l'-1)}{n'(2l'+3)(2l'+2)(2l'+1)2l'(2l'-1)} \quad (3.5.4)$$

$$\Delta E_{ls} = \frac{2Z'^4}{c^2 n'^5} \frac{\left[4l'(l^8+1) - 3 - 12B\right]n'^2 + 4Bl'(l'+1)}{(2l'+3)(2l'+2)(2l'+1)2l'(2l'-1)} \times \begin{cases} l & j = l + \frac{1}{2} \\ -(l+1) & j = l - \frac{1}{2} \end{cases} \quad (3.5.5)$$

$$\Delta E_d = \frac{4BZ'^4}{c^2 n'^5} \frac{3n'^2 - l'(l'+1)}{(2l'+3)(2l'+2)(2l'+1)2l'(2l'-1)} \quad (3.5.6)$$

In above three equations, B is the B in Eq. (3.2.21), i.e.

$$B = \frac{d(d+1) + 2dl}{2} \quad (3.5.7)$$

Let $Z' = Z$, $d = 0$, then Eqs. (3.5.4)–(3.5.6) can be restored to the formulae of hydrogen and hydrogen-like fine structure.

3.6 Calculation of Spin–Orbit Coupling Coefficient

Wang et al. [27] derived formulae for calculating spin–orbit coupling coefficient within WBEPM Theory, and their work will be discussed in the following.
Choose

$$V(r) = -\frac{Z'}{r} + \frac{d(d+1) + 2dl}{2r^2} \quad (3.6.1)$$

To get spin–orbit coupling coefficient, the following formula needs to be calculated:

$$\zeta_{nl} = \langle nl | \xi(r) | nl \rangle \quad (3.6.2)$$

in which

$$\xi(r) = \frac{1}{2c^2 r} \frac{dV(r)}{dr} \quad (3.6.3)$$

After substituting (3.6.1) into (3.6.3), (3.6.2) becomes

$$\zeta_{nl} = \frac{1}{2c^2} \left[Z' \langle nl | r^{-3} | nl \rangle - 2B \langle nl | r^{-4} | nl \rangle \right] \quad (3.6.4)$$

where

$$B = \frac{1}{2}[d(d+1) + 2dl] \quad (3.6.5)$$

Using Hyper-Virial theorem, we get

$$\zeta_{nl} = \frac{Z'^2}{2c^2} \frac{[4l'(l'+1) - 3 - 12B]n'^2 + 4Bl'(l'+1)}{[4l'(l'+1) - 3]l'(l'+1)n'^2} \langle nl|r^{-2}|nl\rangle \quad (3.6.6)$$

Again using

$$\langle nl|r^{-2}|nl\rangle = \left(\frac{2Z'}{n'}\right)\frac{1}{2l'+1}\langle nl|r^{-1}|nl\rangle \quad (3.6.7)$$

and

$$\langle nl|r^{-1}|nl\rangle = \frac{Z'}{n'^2} \quad (3.6.8)$$

Eventually we get

$$\zeta_{nl} = \frac{Z'^4}{c^2 n'^5} \frac{[4l'(l'+1) - 3 - 12B]n'^2 + 4Bl'(l'+1)}{[4l'(l'+1) - 3]l'(l'+1)(2l'+1)}$$

$$(l' > \frac{1}{2}) \quad (3.6.9)$$

If $Z' = Z, d = 0 (n' = n, l' = l)$, the above formula will degenerate to the formula for calculating spin–orbit coupling coefficient of hydrogen and hydrogen-like atoms.

3.7 Relation Between the WBEPM Theory and Slater-Type Orbitals

Slater suggested one-electron potential function $V(r_i)$ of an atomic system to take the following form

$$V(r_i) = -\frac{(Z - s_i)}{r_i} + \frac{n'_i(n'_i - 1)}{2r_i^2} \text{(a.u.)} \quad (3.7.1)$$

Substituting $V(r_i)$ into one-electron Schrodinger equation

$$\left\{-\frac{1}{2}\nabla_i^2 + V(r_i)\right\}\psi_i = \varepsilon_i \psi_i \quad (3.7.2)$$

we can get formulae of Slater-type orbital and one-electron energy ε_i

$$\psi_i = Nr_i^{n'_i - 1} e^{-(Z-s_i)r_i/n'_i} Y_{l,m}(\theta_i, \phi_i) \quad (3.7.3)$$

3.7 Relation Between the WBEPM Theory and Slater-Type Orbitals

and

$$\varepsilon_i = -R(\frac{Z - s_i}{n'_i})^2 \tag{3.7.4}$$

as well as the total energy of electrons for an atomic or molecular system

$$E = \sum_{i=1}^{N} \varepsilon_i = -R \sum_{i=1}^{N} (\frac{Z - s_i}{n'_i})^2 \tag{3.7.5}$$

In above equations, N is normalization factor; $Y_{l,m}(\theta_i, \phi_i)$ is spherical harmonic, which is the same as for hydrogen atom; R is Rydberg constant; n'_i is the effective principle quantum number of i-th electron; s_i is shielding constant of i-th electron owing to shielding by other electrons; $Z - s_i = Z'_i$ is the effective nuclear charge that i-th electron feels.

For n'_i and s_i, Slater devised a set of empirical rules. The basic content of the rules is as follows.

(1) The correspondence between effective principle quantum number n'_i and principle quantum number n_i is

$$\begin{array}{l} n_i \; 1 \; 2 \; 3 \; 4 \; 5 \; 6 \\ n'_i \; 1 \; 2 \; 3 \; 3.7 \; 4.0 \; 4.2 \end{array} \tag{3.7.6}$$

(2) To determine Z'_i, the electrons are arranged into a number of groups in sequential order

$$(1s)(2s, p)(3s, p)(3d)(4s, p)(4d)(4f)(5s, p)(5d) \cdots$$

For an electron in any of those groups, s_i is equal to the sum of the following contributions.

(a) Electron in any group to the right side of the group of interest has no contributions to s_i.
(b) Each other electron within the same group contributes 0.35. (for each other electron in (1s) group, the contribution is 0.3).
(c) If the electron of interest is in (ns,p) group, each electron with principle quantum number (n−1) contributes 0.85, and for each electron with principle quantum number less than (n−1), the contribution is 1.00.
(d) If the electron of interest is in d or f group, each electron in the left side group contributes 1.00.

Slater-type orbitals and the empirical rules have been widely applied to discussions of many physical and chemical problems such as total energy of an atom or an ion, successive ionization energy of an atom, all level limits of atomic X-ray spectral lines, the electronegativity scale, the Hard-Soft-Acid–Base scale, ionic polarizability scale, the law of stability constant of complex, etc., especially to the theoretical calculations of quantum chemistry [21, 30, 32].

Slater-type orbitals are functions without nodal planes. They are completely nonorthogonal. (This can be corrected by a set of orthogonal Slater-type orbitals).

About Slater-type orbitals and the empirical rules, there are two things worth discussing here. First, the potential function that Slater suggested (i.e. Eq. 3.7.1) is based on the assumption that the part $-\frac{Z}{r_i} + \sum_{j,i<j}^{N} \frac{1}{r_{ij}}$ of the Hamiltonian of a many-electron atom can be replaced by the form of Eq. (3.7.1), thus many-electron Hamiltonian $\widehat{H}$ can be written as the sum of one-electron Hamiltonians, and one-electron Schrodinger equation can be obtained, i.e.

$$\widehat{H} = \sum_i \widehat{H}_i \qquad (3.7.7)$$

$$\widehat{H}_i = -\frac{1}{2}\nabla_i^2 + V(r_i) \qquad (3.7.8)$$

$$\widehat{H}_i \psi_i = \left\{ -\frac{1}{2}\nabla_i^2 + V(r_i) \right\} \psi_i = \varepsilon_i \psi_i \qquad (3.7.9)$$

Note that this is just an assumption but not a result derived from the origin of Hamiltonian of a many-electron atom and electronic wave functions; second, the empirical rules were not determined by variation methods but determined by empirical adjustments in order to be consistent with the results of energy levels of stripped atoms, X-ray energy levels, etc.

In Chap. 2, we have expounded that within WBE Theory, the Hamiltonian of the system can be written as the sum of one-electron Hamiltonians of the Weakest Bound Electrons and we have also derived one-electron Schrodinger equation for the Weakest Bound Electron.

For an atomic system, Eq. (2.4.8) and Eq. (2.4.9) are rewritten as

$$\widehat{H}_\mu^0 = -\frac{1}{2}\nabla_\mu^2 + V(r_\mu) \qquad (3.7.10)$$

and

$$\widehat{H}_\mu^0 \psi_\mu^0 = \varepsilon_\mu^0 \psi_\mu^0 \qquad (3.7.11)$$

Equation (3.7.10) represents the formula of non-relativistic one-electron Hamiltonian of the Weakest Bound Electron in the central field approximation. Equation (3.7.11) is the one-electron Schrodinger equation corresponding to $\widehat{H}_\mu^0$.

If $V(r_\mu)$ in Eq. (3.7.10) takes the form of one-electron potential function suggested by Slater, i.e. Equation (3.7.1), then Eqs. (3.7.3), (3.7.4) and (3.7.5) can be obtained.

If $V(r_\mu)$ in Eq. (3.7.10) takes the form of

3.7 Relation Between the WBEPM Theory and Slater-Type Orbitals

Table 3.3 The experimentally measured successive ionization energies (eV) for lithium, carbon, and fluorine atoms

Element	I_1	I_2	I_3	I_4	I_5	I_6	I_7	I_8	I_9
Li	5.392	75.638	122.451						
C	11.260	24.383	47.887	64.492	392.077	489.981			
F	17.422	34.970	62.707	87.138	114.240	157.161	185.182	953.886	1 103.089

$$V(r_\mu) = -\frac{Z'}{r} + \frac{d(d+1) + 2dl}{2r^2} \tag{3.7.12}$$

the related formula in WBEPM Theory can be derived.

Thus, Slater-type orbitals and one-electron wave functions of the Weakest Bound Electrons represented by generalized Laguerre functions within WBEPM Theory are the results at different approximate one-electron potential functions.

About Slate rules, although Slater introduced the methods and principles of choosing n'_i and s_i in reference [29], he did not reveal related details.

In order to help give some kind of deduction, we tried calculations in the following four steps using lithium, carbon and fluorine atoms and ions as examples [33].

Firstly, find experimentally measured successive ionization energies for lithium, carbon and fluorine atoms, which are listed in Table 3.3.

Secondly, for the groups defined by Slater rules, calculate the sum of ionization energies $\sum I_j$ of electrons in each group as well as the average ionization energies over electrons within the group.

$$I_{ave} = \frac{\sum I_j}{m} \tag{3.7.13}$$

where m is the number of electrons within the group.

Thirdly, take Eq. (3.7.4) with ε_i replaced by I_{ave} to calculate s, which is temporally denoted by s_c

Fourthly, use Slater rules to calculate s, which is denoted by s_s. All the results of calculation are listed in Table 3.4.

In Table 3.4, s_c and s_s are very close. Thus, Slater rules include some averageness. In Chaps. 2 and 3, there is

$$-I_\mu \approx \varepsilon_\mu^0 \tag{3.7.13}$$

So we think that s in Slater empirical rules is an approximate average result by treating every electron within the group defined by Slater rules as the Weakest Bound Electron.

Table 3.4 s_c and s_s ($R = 13.6$ eV)

(a) Li

	$j = 1$		$j = 2\,to\,3$		$j = 3$	
$\sum I_j$	5.392		198.089		122.451	
I_{ave}	5.392		99.045		122.451	
S_c	1.741		0.301		~0	
S_s	1.7		0.3		0	

(b) C

	$j = 1\,to\,4$	$j = 2\,to\,4$	$j = 3\,to\,4$	$j = 4$	$j = 5\,to\,6$	$j = 6$
$\sum I_j$	148.022	136.762	112.379	64.492	882.058	489.981
I_{ave}	37.001	45.587	56.190	64.492	441.029	489.981
S_c	2.70	2.338	1.935	1.645	0.305	~0
S_s	2.75	2.4	2.05	1.7	0.3	0

References

1. Zheng NW (1988) A new introduction to atoms. Nanjing Education Press, Nanjing
2. Zheng NW Chinese Science Bulletin., 1977, 22: 531; 1985, 30: 1801; 1986, 31: 1316; 1987, 32: 354. Zheng N W. KEXUE TONGBAO, 1986, 31: 1238; 1987, 32: 1263; 1988, 33: 916.
3. Zheng NW, Wang T, Ma DX, Zhou T, Fan J (2004) Int J Quantum Chem 98:281–290
4. Xu G X, Li L M, Wang D M (1985) Basic principle and Ab Initio method of quantum chemistry, vol 2. Science Press, Beijing
5. Zeng JY (1993) Introduction to quantum mechanics. Peiking University Press, Beijing
6. Yin HJ (1999) Quantum mechanics. University of Science and Technology of China Press, Hefei
7. Slater JC (1981) Quantum theory of atomic structure. Vol. 1 (trans: Yang C H). Shanghai Science and Technology Press, Shanghai
8. (1958). Atomic Physics: Vol. 2 Book 1 (trans: Zhou T Q et al). Higher Education Press, Beijing
9. Zheng NW, Zhou T, Wang T et al (2001) Chin J Chem Phys 14:292
10. Wang LY, Huang JP, Hu GQ (2000) J Nat Sci Hunan Normal Univ 23:39
11. Chen ZD, Chen G (2005) Chin J Chem Phys 18:983
12. Wang ZX, Guo DR (2000) Introduction to special functions. Peiking University Press, Beijing
13. Editing group of mathematical handbook (1979) Mathematical handbook. Higher Education Press, Beijing
14. Wen GW, Wang LY, Wang RD (1990) Chin Sci Bull 35:1231
15. Zhou GL, Lu SC (1995) J Shanghai Normal Univ: Nat Sci Ed 24:40
16. Jia X f (2003) J ShanXi Normal Univ: Nat Sci Ed 20:89
17. Wen GW, Wang RD, Wang LY (1991) Chin Sci Bull 15:1137
18. Zhou GH, Wen GW, Wang LY et al (1994) J At Mol Phys 11:276
19. Chen G (2003) J At Mol Phys 20:89
20. Zeng JY (2001) Quantum mechanics, vol 1, 3rd edn. Science Press, Beijing
21. Xu G X, Li L M (1984) Basic principle and Ab initio method of quantum chemistry, vol 1. Science Press, Beijing
22. Peng HW, Xu XS (1998) Basis of theoretical physics. Peiking University Press, Beijing
23. Levine IN (1982) Quantum chemistry (trans: Ning SG, She JZ, Liu SC). People's Education Press, Beijing
24. Chen CY, Shen HL, Sun GY (1997) Acta Physica Sinica 46:1055

References

25. Chen CY, Zhou RQ, Sun GY (1997) Chin J At Mol Phys 14:657
26. Chen CY, Zhou RQ, Sun GY (1995) Chin J At Mol Phys 12:336
27. Wang LY, Wen GW (1992) Chin Sci Bull 8:708
28. Slater JC (1930) Atomic shielding constants. Phys Rev 36:57–64
29. Pilar FL (1968) Elementary quantum chemistry. McGraw-Hill Book Company, New York
30. Pople JA, Beveridge DL (1976) Approximate molecular orbital theory (trans: Jiang Y). Science Press, Beijing
31. Xu GX, Zhao XZ (1956) Acta Chim Sinica 22:441
32. Foresman JB, Frisch A (1996) Exploring chemistry with electronic structure methods, 2nd edn. Gaussian Inc., Pittsburgh
33. Zheng NW (1992) Univ Chem 7:22

Chapter 4
The Application of the WBE Theory

4.1 Ionization Energy [1–10]

4.1.1 Introduction

The definition of ground-state ionization energy has been given in Refs. [11–14]. The definition includes following information: ① the concept of the Weakest Bound Electron; ② the concept of energy level and ionization limit; ③ the choice of zero-point in quantum chemistry with respect to electronic energy of an atomic or molecular system; ④ total electronic energy of the system is equal to the sum of successive ionization energies. Thus, atomic energy level and ionization energy are one of the important and fundamental properties of chemistry, physics, and astrophysics.

Considering that we not only discuss ground-state ionization energy but also discuss excited-state ionization energy, for the convenience of discussion, here we give a more general definition of ionization energy according to the above definition: the energy required to completely remove one Weakest Bound Electron from a certain energy level of a free particle (atom, molecule), i.e. the energy difference between series limit and energy level, is called ionization energy.

Ionization energy can be measured experimentally (spectroscopy, electron collisions, photoionization, etc.), and the obtained data have been collected in handbook, corpus and database. Atomic successive ionization energies are given in Table 4.1. There are more experimental data for successive ionization energies of the elements with lower atomic number, while for atoms with higher atomic number (Z) and atoms which are highly ionized, the experimental data are relatively less, so there is a large gap to fill in. This is related to difficult experimental preparation and complicated structure of energy level. However, these data are exactly what the fields such as space technology, nuclear fusion, laser technology, etc. are interested in, so it is the hot topic of the current experimental and theoretical studies.

Many theoretical methods have been used for the calculation of atomic energy level and ionization energy (including bound-state and excited-state ionization

Table 4.1 Atomic successive ionization energies (eV)

Neutral atoms to +7 ions

Z	Element	I	II	III	IV	V	VI	VII	VIII
1	H	13.598 443							
2	He	24.587 387	54.417 760						
3	Li	5.391 719	75.640 0	122.454 29					
4	Be	9.322 70	18.211 14	153.896 61	217.718 65				
5	B	8.298 02	25.154 8	37.930 64	259.375 21	340.225 80			
6	C	11.260 30	24.383 3	47.887 8	64.493 9	392.087	489.993 34		
7	N	14.534 1	29.6013	47.449 24	77.473 5	97.890 2	552.0718	667.046	
8	O	13.618 05	35.121 1	54.935 5	77.413 53	113.899 0	138.119 7	739.29	871.410 1
9	F	17.422 8	34.970 8	62.708 4	87.139 8	114.242 8	157.1651	185.186	953.911 2
10	Ne	21.564 54	40.962 96	63.45	97.12	126.21	157.93	207.275 9	239.098 9
11	Na	5.139 075	47.286 4	71.620 0	98.91	138.40	172.18	208.50	264.25
12	Mg	7.646 235	15.035 27	80.143 7	109.265 5	141.27	186.76	225.02	265.96
13	Al	5.985 763	18.828 55	28.447 65	119.992	153.825	190.49	241.76	284.66
14	Si	8.151 68	16.345 84	33.493 02	45.141 81	166.767	205.27	246.5	303.54
15	P	10.486 69	19.769 5	30.202 7	51.443 9	65.025 1	220.421	263.57	309.60
16	S	10.360 01	23.337 88	34.79	47.222	72.594 5	88.053 0	280.948	328.75
17	Cl	12.967 63	23.813 6	39.61	53.465 2	67.8	97.03	114.195 8	348.28
18	Ar	15.759 610	27.629 66	40.74	59.81	75.02	91.009	124.323	143.460
19	K	4.340 663 3	31.63	45.806	60.91	82.66	99.4	117.56	154.88
20	Ca	6.113 16	11.871 72	50.913 1	67.27	84.50	108.78	127.2	147.24
21	Sc	6.561 49	12.799 77	24.756 66	73.489 4	91.65	110.68	138.0	158.1
22	Ti	6.828 12	13.575 5	27.4917	43.267 2	99.30	119.53	140.8	170.4

(continued)

4.1 Ionization Energy

Table 4.1 (continued)

Neutral atoms to +7 ions

Z	Element	I	II	III	IV	V	VI	VII	VIII
23	V	6.746 19	14.618	29.311	46.709	65.281 7	128.13	150.6	173.4
24	Cr	6.766 51	16.485 7	30.96	49.16	69.46	90.634 9	160.18	184.7
25	Mn	7.434 02	15.640 0	33.668	51.2	72.4	95.6	119.203	194.5
26	Fe	7.902 4	16.187 7	30.652	54.8	75.0	99.1	124.98	151.06
27	Co	7.881 01	17.084	33.50	51.3	79.5	102.0	128.9	157.8
28	Ni	7.639 8	18.168 84	35.19	54.9	76.06	108	133	162
29	Cu	7.726 38	20.292 4	36.841	57.38	79.8	103	139	166
30	Zn	9.394 199	17.964 39	39.723	59.4	82.6	108	134	174
31	Ga	5.999 301	20.515 14	30.71	64				
32	Ge	7.899 43	15.934 61	34.224 1	45.713 1	93.5			
33	As	9.788 6	18.589 2	28.351	50.13	62.63	127.6		
34	Se	9.752 39	21.19	30.820 4	42.945 0	68.3	81.7	155.4	
35	Br	11.813 8	21.591	36	47.3	59.7	88.6	103.0	192.8
36	Kr	13.999 61	24.359 84	36.950	52.5	64.7	78.5	111.0	125.802
37	Rb	4.177 128	27.289 5	40	52.6	71.0	84.4	99.2	136
38	Sr	5.694 85	11.030 1	42.89	57	71.6	90.8	106	122.3
39	Y	6.217 3	12.224	20.52	60.597	77.0	93.0	116	129
40	Zr	6.633 90	13.1	22.99	34.34	80.348			
41	Nb	6.758 85	14.0	25.04	38.3	50.55	102.057	125	

+8 ions to +15 ions

Z	Element	IX	X	XI	XII	XIII	XIV	XV	XVI
9	F	1 103.117 6							
10	Ne	1 195.828 6	1 362.199 5						

(continued)

Table 4.1 (continued)

+8 ions to +15 ions

Z	Element	IX	X	XI	XII	XIII	XIV	XV	XVI
11	Na	299.864	1 465.121	1 648.702					
12	Mg	328.06	367.50	1 761.805	1 962.665 0				
13	Al	330.13	398.75	442.00	2 085.98	2 304.141 0			
14	Si	351.12	401.37	476.36	523.42	2 437.63	2 673.182		
15	P	372.13	424.4	479.46	560.8	611.74	2 816.91	3 069.842	
16	S	379.55	447.5	504.8	564.44	652.2	707.01	3 223.78	3 494.189 2
17	Cl	400.06	455.63	529.28	591.99	656.71	749.76	809.40	3 658.521
18	Ar	422.45	478.69	538.96	618.26	686.10	755.74	854.77	918.03
19	K	175.817 4	503.8	564.7	629.4	714.6	786.6	861.1	968
20	Ca	188.54	211.275	591.9	657.2	726.6	817.6	894.5	974
21	Sc	180.03	225.18	249.798	687.36	756.7	830.8	927.5	1 009
22	Ti	192.1	215.92	265.07	291.500	787.84	863.1	941.9	1 044
23	V	205.8	230.5	255.7	308.1	336.277	896.0	976	1 060
24	Cr	209.3	244.4	270.8	298.0	354.8	384.168	1 010.6	1 097
25	Mn	221.8	248.3	286.0	314.4	343.6	403.0	435.163	1 134.7
26	Fe	233.6	262.1	290.2	330.8	361.0	392.2	457	489.256
27	Co	186.13	275.4	305	336	379	411	444	511.96
28	Ni	193	224.6	321.0	352	384	430	464	499
29	Cu	199	232	265.3	369	401	435	484	520
30	Zn	203	238	274	310.8	419.7	454	490	542

(continued)

4.1 Ionization Energy

Table 4.1 (continued)

+8 ions to +15 ions

Z	Element	IX	X	XI	XII	XIII	XIV	XV	XVI
36	Kr	230.85	268.2	308	350	391	447	492	541
37	Rb	150	277.1						
38	Sr	162	177	324.1					
39	Y	146.2	191	206	374.0				
42	Mo	164.12	186.4	209.3	230.28	279.1	302.60	544.0	570

+16 ions to +23 ions

Z	Element	XVII	XVIII	XIX	XX	XXI	XXII	XXIII	XXIV
17	Cl	3 946.296 0							
18	Ar	4 120.885 7	4 426.229 6						
19	K	1 033.4	4 610.8	4 934.046					
20	Ca	1 087	1 157.8	5 128.8	5 469.864				
21	Sc	1 094	1 213	1 287.97	5 674.8	6 033.712			
22	Ti	1 131	1 221	1 346	1 425.4	6 249.0	6 625.82		
23	V	1 168	1 260	1 355	1 486	1 569.6	6 851.3	7 246.12	
24	Cr	1 185	1 299	1 396	1 496	1 634	1 721.4	7 481.7	7 894.81
25	Mn	1 224	1 317	1 437	1 539	1 644	1 788	1 879.9	8 140.6
26	Fe	1 266	1 358	1 456	1 582	1 689	1 799	1 950	2 023
27	Co	546.58	1 397.2	1 504.6	1 603	1 735	1 846	1 962	2 119
28	Ni	571.08	607.06	1 541	1 648	1 756	1 894	2 011	2 131
29	Cu	557	633	670.588	1 697	1 804	1 916	2 060	2 182
30	Zn	579	619	698	738	1 856			
36	Kr	592	641	786	833	884	937	998	1 051
42	Mo	636	702	767	833	902	968	1 020	1 082

(continued)

Table 4.1 (continued)

+24 ions to +29 ions

Z	Element	XXV	XXVI	XXVII	XXVIII	XXIX	XXX
25	Mn	8 571.94					
26	Fe	8 828	9 277.69				
27	Co	2 219.0	9 544.1	10 012.12			
28	Ni	2 295	2 399.2	10 288.8	10 775.40		
29	Cu	2 308	2 478	2 587.5	11 062.38	11 567.617	
36	Kr	1 151	1 205.3	2 928	3 070	3 227	3 381
42	Mo	1 263	1 323	1 387	1 449	1 535	1 601

Data source Lide DR. CRC handbook of chemistry and physics, 86th edn. Taylor & Francis, New York, 2005–2006

4.1 Ionization Energy

energy), including Relativistic Configuration Interaction (RCI), Relativistic Many-Body Perturbation Theory (RMBPT), Relativistic Coupled Cluster (RCC), Multi-Configuration Dirac–Fock method (MCDF), R-matrix method, Density Functional Theory (DFT), Weakest Bound Electron Potential Model Theory (WBEPM Theory), etc. [1–10, 15–30]. For instance, Jursic [30] calculated the first to the fourth ionization energies for period 2 elements such as carbon, nitrogen, oxygen and fluorine by using ab initio method (HF, MP2, MP3, MP4DQ, QCISD, G1, G2 and G2MP2) and Density Functional Theory (DFT) method (B3LYP, B3P86, B3PW91, XALW91, XAWPHA, HFS, HFB, BLYP, BP86, BPW91, BVWN, XALP86, XAPW91, XAVWN, SLYP, SP86, SPW91 and SVWN). Large Gaussian 6-311 + + G(3df,3pd) basis set has been used in the calculations. And in the paper, comparisons between calculated values and experimental values were given and the applicability of these methods was evaluated.

From Tables 4.2, 4.3, 4.4, and 4.5, roughly we can tell the precision that each method currently can reach.

It is not hard to find that WBEPM Theory is quite accurate for the calculations of atomic ionization energy (ground-state and excited-state).

Table 4.2 Comparison of ionization energies calculated by MCHF method and WBEPM theory with experimental values for oxygen-like sequences $1s^2 2s^2 2p^3(^4S^0)3s^3S_1^0$ (eV)

Z	I_{exp}	I_{calc}^a (MCHF)	$I_{exp} - I_{calc}^a$	I_{calc}^b (WBEPM theory)	$I_{exp} - I_{calc}^b$
8	4.097	3.964	+0.133	4.095	+0.002
9	12.299	12.160	+0.139	12.318	−0.019
10	23.849	23.702	+0.147	23.827	+0.022
11	38.578	38.390	+0.188	38.539	+0.039
12	56.398	56.196	+0.202	56.383	+0.015
13	77.265	77.094	+0.171	77.304	−0.039
14	101.113	101.071	+0.042	101.259	−0.146
15	128.294	128.120	+0.174	128.224	+0.070
16	158.471	158.240	+0.231	158.186	+0.285
17	190.961	191.428	−0.467	191.148	−0.187
18	–	–	–	227.128	–
19	266.145	–	–	266.158	−0.013
20	308.274	–	–	308.283	−0.009
21	353.665	–	–	353.567	+0.098
22	402.069	–	–	402.083	−0.014

Data source I_{exp}, I_{calc}^a and I_{calc}^b in the table were taken from (1) Fuhr JR, Martin WC, Musgrove A, Sugar J, Wiese WL. NIST Atomic Spectroscopic Database, Version 2.0, 1996; URL: http://physics.nist.gov/PhysRefData. (2) Lindgard A, Nielsem SE. J Phys B, 1975, 8:1183. (3) Zhang NW, Wang T, Ma DX, et al. Int J Quantum Chem, 2004, 98:281–290 (Table IV)

Table 4.3 Comparison of the first to the fourth ionization energies of nitrogen calculated by ab initio and DFT method with experimental values (eV)

Theory	I_1	I_2	I_3	I_4
HF	13.91	29.19	47.32	75.03
MP2	14.59	29.60	47.49	76.37
MP3	14.56	29.57	47.41	76.89
MP4DQ	14.51	29.51	47.32	77.14
QCISD	14.45	29.45	47.24	77.36
QCISD(T)	14.47	29.46	47.27	77.36
G1	14.47	29.44	47.23	77.52
G2	14.48	29.46	47.27	77.49
G2MP2	14.43	29.44	47.25	77.53
B3LYP	14.67	29.99	48.34	76.60
B3P86	15.30	30.49	48.65	77.09
B3PW91	14.78	30.00	48.17	76.48
HFS	13.98	28.82	46.68	74.09
HFB	14.01	29.16	47.42	75.23
XALPHA	14.51	29.47	47.22	74.67
BLYP	14.51	29.83	48.24	76.46
BP86	14.77	29.91	48.08	76.57
BPW91	14.75	29.34	48.15	76.44
BVWN	15.51	30.86	49.31	77.86
XALYP	15.00	30.14	48.05	75.90
XAP86	15.26	30.21	47.89	76.01
XAPW91	15.24	30.24	47.95	75.88
XAVWN	16.01	31.16	49.11	77.30
SLYP	14.55	29.41	47.30	75.32
SP86	14.81	29.49	47.15	75.42
SPW91	14.78	29.52	47.21	75.29
SVWN	15.47	30.51	48.36	76.71
Exp. value	14.54	29.59	47.43	77.45

Data source Jursic BS. Int J Quantum Chem, 1997, 64:255 (Table III)

4.1.2 Iso-spectrum-level Series and the Differential Law of Ionization Energy in the Series

In the past, people always discussed the law of the change of ionization energy with the nuclear charge in the concept of isoelectronic sequence. The so-called isoelectronic sequence is a sequence of atoms and ions which contain the same number of electrons and sequentially increased nuclear charge. For instance, Ar I,

4.1 Ionization Energy

Table 4.4 Comparison of ground-state remove energies given by V^{N-M} approximations with experimental values for KrVIII to KrI (a.u.)[①]

Species	State		Exp. value	Calculations
Kr VIII	4s	$^2S_{1/2}$	−4.623 17	−4.626 99
Kr VII	$4s^2$	$^1S_0$	−8.702 47	−8.640 60
Kr VI	$4s^24p$	$^2P^0_{1/2}$	−11.587 09	−11.524 81
Kr V	$4s^24p^2$	$^3P_0$	−13.964 59	−13.890 50
Kr IV	$4s^24p^3$	$^4S^0_{3/2}$	−15.893 75	−15.747 36
Kr III	$4s^24p^4$	$^3P_2$	−17.251 63	−17.039 29
Kr II	$4s^24p^5$	$^2P^0_{3/2}$	−18.146 84	−17.883 92
Kr I	$4s^24p^6$	$^1S_0$	−18.661 32	−18.287 61

Data source Dzuba VA. Phys Rev A, 2005, 71:032512 (Table III). Comparison of calculated values with experimental values is shown in this table for the energies required to remove all valence electrons from a ground state. For all ions and neutral atoms, the precisions of calculations are very close with relative errors roughly smaller than 2%

Notes [①] V^{N-M} is a good starting point for calculations using RCI

Table 4.5 Two-electron energies of the Ba ground state in different approximations (a.u.)

Theory	E_{exp}[⑤]	E_{calc}	$E_{exp} - E_{calc}$	$(E_{exp} - E_{calc})/E_{exp}$ (%)
RHF[①]	−0.559 15	−0.504 02	−0.055 13	+9.86
MBPT[②]	−0.559 15	−0.540 53	−0.018 62	+3.33
CI[③]	−0.559 15	−0.527 90	−0.031 25	+5.59
CI + MBPT[④]	−0.559 15	−0.560 65	$+1.5 \times 10^{-3}$	−0.26

Data source Dzuba VA, Johnson WR. Phys Rev A, 1998, 57:2459 (Table II)

Notes [①] Single-configuration approximation. No correlations are included
[②] Single-configuration approximation. Core-valence correlations are included
[③] Standard CI method. Only valence-valence correlations are included
[④] CI + MBPT method. Core-valence and valence-valence correlations are included
[⑤] This is the sum of the ionization energies of Ba and Ba$^+$

K II, Ca III, Sc IV, ... comprise the Ar I isoelectronic sequence. The concept of isoelectronic sequence only provides information about electronic configurations. It is well known that under electronic configuration, there are spectral terms and under spectral term, there are energy levels. For instance, the ground-state electronic configuration of C atom is [He]$2s^22p^2$. There are three spectral terms under this electronic configuration, i.e., ^3P, ^1D, and ^1S. These spectral terms further generate energy levels $^3P_2$, $^3P_1$, $^3P_0$, $^1D_2$, and $^1S_0$. The energy difference between energy levels is small within a spectral term, but under an electronic configuration, the energy difference between spectral terms is remarkable. Under C I ground-state electronic configuration, the energy difference between the highest energy level $^1S_0$ and the lowest energy level $^3P_0$ can be 21648.4 cm^{-1}. The situation is roughly the same for excited-state electronic configurations. The concept of isoelectronic sequence doesn't provide related information of spectral term and energy level. However, why can the concept of isoelectronic sequence be used for the study of the law of ground-state ionization energy? The situation for ground state is somewhat special. The so-called special is that "ground-state" conditions have been conventionally attached to the concept of isoelectronic sequence. This actually means that problems are solved in the concept of the lowest energy level of sequence members. Thus, overall, the concept of isoelectronic sequence is crude for discussion of the law of ionization energy, and more refined concepts are required to seek. Therefore, we proposed the concept and definition of iso-spectrum-level series.

The so-called iso-spectrum-level series is a series of energy levels which have the same level symbol under a given electronic configuration in an isoelectronic sequence. For instance, C I [He]$2s^22p3s$ $^3P_2^0$, N II [He]$2s^22p3s$ $^3P_2^0$, O III [He]$2s^22p3s$ $^3P_2^0$, F IV [He]$2s^22p3s$ $^3P_2^0$, Ne V [He]$2s^22p3s$ $^3P_2^0$, ..., comprise an iso-spectrum-level series named C I [He]$2s^22p3s$ $^3P_2^0$. C I [He]$2s^22p^2$ $^1S_0$, N II [He]$2s^22p^2$ $^1S_0$, O III [He]$2s^22p^2$ $^1S_0$, F IV [He]$2s^22p^2$ $^1S_0$, Ne V [He]$2s^22p^2$ $^1S_0$, ..., make up C I [He]$2s^22p^2$ $^1S_0$ iso-spectrum-level series; Fe I [Ar]$3d^64s^2$ $^5D_4$, Co II [Ar]$3d^64s^2$ $^5D_4$, Ni III [Ar]$3d^64s^2$ $^5D_4$, Cu IV [Ar]$3d^64s^2$ $^5D_4$, Zn V [Ar]$3d^64s^2$ $^5D_4$, ..., comprise Fe I [Ar]$3d^64s^2$ $^5D_4$ iso-spectrum-level series. From above examples, we can see that the symbol to represent iso-spectrum-level series consists of three parts: the beginning is the symbol of the element and ionization status for a member of a given isoelectronic sequence, such as C I, N II, O III, etc.; the middle part is the given electronic configuration, such as $2s^22p3s$; the end is the energy level symbol such as $^3P_2^0$. The concept of iso-spectrum-level series brings people from the level of electronic configuration to the level of energy level. In a given iso-spectrum-level series, electron configuration, spectral term, and spectrum energy level are identical, and the sole variable is the nuclear charge Z. Thus, the ionization energy can be considered as a univariant function of nuclear charge Z, which makes it more convenient to study the law of the change of ionization energy.

Let's restate the general definition of ionization energy which was described previously: the energy required to completely remove one Weakest Bound Electron from a certain energy level of a free particle (atom, molecule), i.e. the energy difference between series limit and energy level, is called ionization energy.

According to this definition, we have

$$I_{exp} = -\varepsilon_\mu = T_{limit} - T(n) \tag{4.1.1}$$

T_{limit} is the series limit, $T(n)$ is the energy of a given energy level in the series, and both T_{limit} and $T(n)$ are determined relative to the ground-state energy level. ε_μ is the true energy of the Weakest Bound Electron at a given energy level.

Now let's discuss the law of the change of ionization energy with the nuclear charge in a given iso-spectrum-level series.

For a given iso-spectrum-level series, the sole variable is nuclear charge Z. In Eq. (4.1.1), all quantities change with Z, then

4.1 Ionization Energy

$$I_{exp}(Z) = T_{limit}(Z) - T(Z, n) \tag{4.1.2}$$

Z and n in the parenthesis represent which element and the principle quantum number of the Weakest Bound Electron at a given energy level, respectively.

Take the first difference of ionization energy:

$$\begin{aligned}\Delta I_1(Z+1, Z) &= I_{exp}(Z+1) - I_{exp}(Z) \\ &= T_{limit}(Z+1) - T(Z+1, n) \\ &\quad - [T_{limit}(Z) - T(Z, n)]\end{aligned} \tag{4.1.3}$$

and

$$\begin{aligned}\Delta I_2(Z+2, Z+1) &= I_{exp}(Z+2) - I_{exp}(Z+1) \\ &= T_{limit}(Z+2) - T(Z+2, n) \\ &\quad - [T_{limit}(Z+1) - T(Z+1, n)]\end{aligned} \tag{4.1.4}$$

Furtherly, take the second difference of ionization energy:

$$\begin{aligned}\Delta^2 I &= \Delta I_2(Z+2, Z+1) - \Delta I_1(Z+1, Z) \\ &= [T_{limit}(Z+2) - 2T_{limit}(Z+1) + T_{limit}(Z)] \\ &\quad - [T(Z+2, n) - 2T(Z+1, n) + T(Z, n)]\end{aligned} \tag{4.1.5}$$

We found that on a finite interval, there is a good linear relationship between the first difference of ionization energy ΔI and Z along an iso-spectrum-level series; the second difference $\Delta^2 I$ is close to a constant. This is called the differential law of ionization energy. Many examples about the change of the ground-state ionization energy with Z have been given in Ref. [1], and readers can refer to it if they are interested. Here, we gave several more examples of the excited state and ground state to make this clear.

Examples: B I [He]$2s^2 2p$ $^2P^0_{1/2}$ (Table 4.6); B I [He]$2s^2 3s$ $^2S_{1/2}$ (Table 4.7); B I [He]$2s^2 3p$ $^2P^0_{3/2}$ (Table 4.8); B I [He]$2s^2 3d$ $^2D_{5/2}$ (Table 4.9); B I [He]$2s^2 2p^2$ $^4P_{1/2}$ (Table 4.10); B I [He]$2p^3$ $^2D_{3/2}$ (Table 4.11); and Fig. 4.1.

In Chap. 2, we have shown that the electronic Hamiltonian of an atomic or molecular system can be written as the sum of one-electron Hamiltonians of the Weakest Bound Electrons. And the one-electron Hamiltonian of the Weakest Bound Electron is equal to the sum of non-relativistic Hamiltonian and relativistic Hamiltonian. In the Breit-Pauli approximation, the one-electron Hamiltonian of the Weakest Bound Electron can be further simplified to be the sum of non-relativistic part and relativistic part (i.e., mass-velocity term, self spin-orbital coupling term of the Weakest Bound Electron μ, Darwin term). Therefore, the experimental ionization energy $I_{exp}(Z)$ can be approximately written as the sum of non-relativistic part $I_{nr}(Z)$ and relativistic correction part $I_r(Z)$ as well. Thus, one has

Table 4.6 The differences of ionization energy along B I [He]$2s^2 2p$ $^2P^0_{1/2}$ iso-spectrum-level series (eV)

Z for each series member	Series limit[1] (cm^{-1})	I_{exp}	ΔI_{exp}	$\Delta^2 I_{exp}$
5	66 928.10	8.298		
6	196 664.7	24.384	16.086	
7	382 703.8	47.450	23.066	6.98
8	624 382.0	77.414	29.964	6.90
9	921 430.0	114.244	36.830	6.87
10	1 273 781[2]	157.930	43.686	6.856
11	1 681 700	208.506	50.576	6.890
12	2 145 100	265.961	57.457	6.881
13	2 662 650	330.129	64.168	6.711
14	3 237 300	401.377	71.248	7.080
15	3 867 100	479.463	78.086	6.838
16	4 552 500	564.443	84.980	6.894
17	5 293 800	656.353	91.911	6.931
18	6 091 100	755.206	98.853	6.942

Notes [1] Data were taken from http://physics.nist.gov; 1 eV = 8065.469 cm^{-1}
[2] Due to the possible error in the data from http://physics.nist.gov, we instead take it from: Weast RC. Handbook of chemistry and physics, 62nd edn. CRC Press, Inc., Boca Raton, Florida, E-65 (1981–1982)

Table 4.7 The differences of ionization energy along B I [He]$2s^2 3s$ $^2S_{1/2}$ iso-spectrum-level series (eV)

Z for each series member	I_{exp}[1]	ΔI_{exp}	$\Delta^2 I_{exp}$
5	3.334		
6	9.934	6.6	
7	20.011	10.077	3.477
8	33.075	13.064	2.987
9	49.182	16.107	3.043
10	68.339	19.157	3.05
11	90.552	22.213	3.056
12	115.852	25.300	3.087
13	144.024	28.172	2.872

Notes [1] Zheng NW, Wang T. Int J Quantum Chem, 2004, 98:495–501 (Table II)

$$I_{exp}(Z) \approx I_{nr}(Z) + I_r(Z) \qquad (4.1.6)$$

$I_{nr}(Z)$ and $I_r(Z)$ have two different properties as follows: ① I_{nr} makes much more contribution to I_{exp} than I_r does. ② As we know, the repulsive energies between electrons scale as r^{-1}, so go as Z. The electron-nuclear Coulomb attractive energies

4.1 Ionization Energy

Table 4.8 The differences of ionization energy along B I [He]$2s^2 3p\ ^2P^0_{3/2}$ iso-spectrum-level series (eV)

Z for each series member	I_{exp} [1]	ΔI_{exp}	$\Delta^2 I_{exp}$
5	2.271		
6	8.050	5.779	
7	16.986	8.936	3.157
8	29.029	12.043	3.107
9	44.124	15.095	3.052
10	62.270	18.146	3.051
11	83.476	21.206	3.060

Notes [1] Zheng NW, Wang T. Int J Quantum Chem, 2004, 98:495–501 (Table II)

Table 4.9 The differences of ionization energy along B I [He]$2s^2 3d\ ^2D_{5/2}$ iso-spectrum-level series (eV)

Z for each series member	I_{exp} [1]	ΔI_{exp}	$\Delta^2 I_{exp}$
5	1.508		
6	6.337	4.829	
7	14.315	7.978	3.149
8	25.396	11.081	3.103
9	39.540	14.144	3.063
10	56.710	17.170	3.026
11	76.994	20.284	3.114
12	100.312	23.318	3.034
13	126.497	26.185	2.867
14	155.919	29.422	3.237
15	188.331	32.412	2.99
16	223.717	35.386	2.974
17	262.078	38.361	2.975

Notes [1] Zheng NW, Wang T. Int J Quantum Chem, 2004, 98:495–501 (Table II)

scale as $-\frac{2Z}{r}$, then go as Z^2. The kinetic energies of electron scale as r^{-2}, so also go as Z^2. Thus, $I_{nr}(Z)$ is a function of Z with Z^2 as the highest power. However, the mass-velocity term, the spin-orbital coupling term and Darwin term increase as the fourth power of Z. Thus, I_r is a function of Z^4 [7–9, 12, 15, 31, 32].

From Eq. (4.1.6), we can get

$$\Delta I_{exp}(Z+1, Z) = \Delta I_{nr}(Z+1, Z) + \Delta I_r(Z+1, Z) \quad (4.1.7)$$

In a given iso-spectrum-level series, the difference of relativistic ionization energies between neighboring members, i.e., $\Delta I_r(Z+1, Z)$ is very small, so it can be omitted temporally. Then

Table 4.10 The differences of ionization energy along B I [He]$2s^2 2p^2$ $^4P_{1/2}$ iso-spectrum-level series (eV)

Z for each series member	Series limit[1] (cm^{-1})	Energy level[2] (cm^{-1})	I_{exp}	ΔI_{exp}	$\Delta^2 I_{exp}$
5	66 928.10	28 805	4.727		
6	196 664.7	43 000	19.052	14.325	
7	382 703.8	57 187	40.359	21.307	6.982
8	624 382	71 177	68.589	28.230	6.923
9	921 430	86 035	103.577	34.988	6.758
10	1 273 781[3]	99 030	145.652	42.075	7.087
11	1 681 700	14 978	194.250	48.598	6.523
12	2 145 100	29 890	249.856	55.606	7.008
13	2 662 650	44 420	312.223	62.367	6.761
14	3 237 300	61 010	381.414	69.191	6.824
15	3 867 100	77 177	457.496	76.08	6.891
16	4 552 500	93 882	540.404	82.908	6.828

Notes [1] http://physics.nist.gov
[2] Safronova MS, Johnson WR, Safronova UI. Phys Rev A, 1996, 54:2850–2862 (Table IV)
[3] Weast RC. Handbook of chemistry and physics, 62nd edn. CRC Press, Inc., Boca Raton, Florida, E-65 (1981–1982)

Table 4.11 The differences of ionization energy along B I [He]$2p^3$ $^2D_{3/2}$ iso-spectrum-level series (eV)

Z for each series member	Series limit (cm^{-1})	Energy level (cm^{-1})	I_{exp}	ΔI_{exp}	$\Delta^2 I_{exp}$
6	196 664.7	150 468	5.728		
7	382 703.8	203 089	22.270	16.542	
8	624 382	255 186	45.775	23.505	6.923
9	921 430	307 273	76.146	30.371	6.866
10	1 273 781[1]	359 601	113.344	37.199	6.828
11	1 681 700	412 395	157.375	44.031	6.832
12	2 145 100	465 818	208.206	50.831	6.800
13	2 662 650	520 140	265.640	57.433	6.603
14	3 237 300	575 450	330.030	64.390	6.957
15	3 867 100	631 961	401.109	71.079	6.689
16	4 552 500	689 910	478.904	77.795	6.716

Data source Safronova MS, Johnson WR, Safronova UI. Phys Rev A, 1996, 54:2850–2862 (Table IV)

Notes [1] Weast RC. Handbook of chemistry and physics, 62nd edn. CRC Press, Inc., Boca Raton, Florida, E-65 (1981–1982)

4.1 Ionization Energy

Fig. 4.1 The plot of the first differences of experimental ionization energy ΔI_{exp} versus the nuclear charge Z in a carbon-like sequence. (Reproduced from Zheng NW, Wang T (2003) Chem Phys Lett 376:557–565 (Fig. 1)). (1) C I [He]$2s^2 2p^2$ $^3P_2$ series; (2) C I [He]$2s^2 2p3s$ $^3P_2^0$ series; (3) C I [He]$2s^2 2p3d$ $^1F_3^0$ series; (4) C I [He]$2s^2 2p4d$ $^3D_3^0$ series

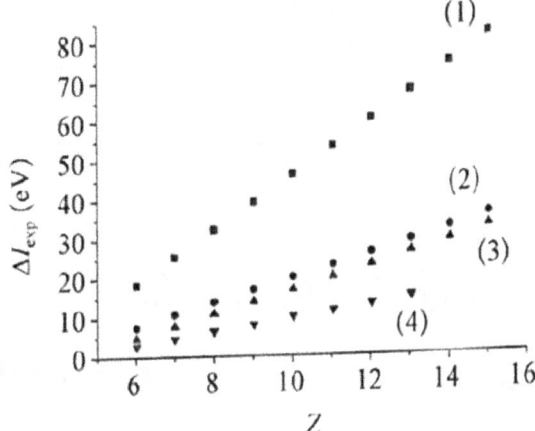

$$\Delta I_{exp}(Z+1, Z) \approx \Delta I_{nr}(Z+1, Z) \quad (4.1.8)$$

Further one has

$$\Delta^2 I_{exp} \approx \Delta^2 I_{nr} \quad (4.1.9)$$

In a given iso-spectrum-level series, $\Delta^2 I_{exp} \approx$ constant because $\Delta Z = 1$, and by using the criteria that variate difference method applies to determine the type of a function, we can immediately conclude that the relation between the experimental ionization energy I_{exp} and the nuclear charge Z is approximately parabolic.

Rewrite several important formulae of WBEPM Theory that were given previously

$$E_{el} = -\sum_{\mu=1}^{N} I_\mu \quad (4.1.10)$$

$$E_{el} = \sum_{\mu=1}^{N} \varepsilon_\mu \quad (4.1.11)$$

$$\varepsilon = -\frac{RZ'^2}{n'^2} \approx \varepsilon_\mu^0 \quad (4.1.12)$$

or

$$\varepsilon = -\frac{Z'^2}{2n'^2}(\text{a.u.}) \approx \varepsilon_\mu^0 \quad (4.1.13)$$

and

$$I_{exp} = -\varepsilon_\mu \quad (4.1.14)$$

Obviously, one has

$$\varepsilon_\mu = \varepsilon_\mu^0 + \Delta E_c + \Delta E_r \qquad (4.1.15)$$

where ε_μ^0 is the non-relativistic one-electron energy of the weakest bound electron in the central field approximation. The correlation energy ΔE_c and relativistic correction must be added so that the sum will equal the true energy ε_μ of the weakest bound electron at a given energy level.

Combining Eqs. (4.1.6), (4.1.14), and (4.1.15) gives

$$I_{exp} = -\varepsilon_\mu = -\left[\left(\varepsilon_\mu^0 + \Delta E_c\right) + \Delta E_r\right] \approx I_{nr} + I_r \qquad (4.1.16)$$

By neglecting relativistic effect, the above equation becomes

$$I_{exp} \approx I_{nr} \approx -\left(\varepsilon_\mu^0 + \Delta E_c\right) \qquad (4.1.17)$$

If we neglect again ΔE_c on the right side of above equation and substitute Eqs. (4.1.1) and (4.1.13) into it, we get

$$I_{exp} = T_{limit} - T(n) \approx I_{nr} \approx \frac{RZ'^2}{n'^2} \qquad (4.1.18)$$

We have previously indicated that in a given iso-spectrum-level series, the relation between ionization energy and the nuclear charge Z is approximately parabolic based on differential law, and based on our studies, we give the following detailed analytical formula for this parabolic relation between I and Z.

$$\begin{aligned} I_{exp} = T_{limit} - T(n) &\approx I_{nr} \approx \frac{RZ'^2}{n'^2} \\ &= \frac{R}{n'^2}[(Z-\sigma)^2 + g(Z-Z_0)] \end{aligned} \qquad (4.1.19)$$

where n' is the effective principle quantum number, and $n' = n + d$ as mentioned in Chap. 3. d is only used to adjust the integral quantum number n and l to be nonintegral n' and l' for the weakest bound electron, so it is basically unrelated to Z. Thus, for a given series, n' is a constant, σ is the screening constant of the first element of the series, Z_0 is the nuclear charge of the first element, and the parameter g is called relatively increase factor, which indicates the contribution to the effective nuclear charge due to the increase of the nuclear charge in the series, which in fact shows that the nuclear charge can't be completely shielded by electrons. It is worth to point out that the first member in a given series is generally assigned as the first element. However, this assignment is random rather than absolute, especially in the case that in order to improve the accuracy of calculations, we divide Z into segments, i.e., Z[a, b], and the interval *ab* shouldn't be too large in order to keep the good linear

4.1 Ionization Energy

relationship between the first difference of ionization energy and the nuclear charge Z.

The method to determine n', σ and g has been given in Refs. [1, 4, 5]. Its procedure is as follows.

(1) From (4.1.5) and (4.1.19), we get

$$\Delta^2 I_{exp} \approx \frac{2R}{n'^2} \tag{4.1.20}$$

or

$$\Delta^2 I_{exp} = [T_{limit}(Z+2) - 2T_{limit}(Z+1) + T_{limit}(Z)]$$
$$- [T(Z+2, n) - 2T(Z+1, n) + T(Z, n)]$$
$$\approx \frac{2R}{n'^2} \tag{4.1.21}$$

By plotting ΔI_{exp} versus Z, we can also get

$$\text{tg}\alpha \approx \frac{2R}{n'^2} \tag{4.1.22}$$

Because $\Delta^2 I_{exp}$ is approximately a constant, by average we can get n' from Eq. (4.1.20) or from the linear slope $\text{tg}\alpha$ of $\Delta I_{exp} \sim Z$.

(2) For the assigned first element, since $Z = Z_0$, $g(Z - Z_0) = 0$, we have

$$I_{exp} \approx \frac{R}{n'^2}(Z - \sigma)^2 \tag{4.1.23}$$

With I_{exp} and n', we can obtain σ.

(3) We get the average value of g with I_{exp} and n' of several members as well as σ.

After n', σ, and g have been determined, they are substituted into Eq. (4.1.19) to calculate the value of I_{nr}.

The contribution of I_r to I_{exp} was initially neglected as mentioned previously. Now it is time to further consider this correction. For a given iso-spectrum-level series, the relativistic corrections in the Breit-Pauli approximation can be represented by a fourth-order polynomial in the nuclear charge Z, i.e.

$$I_r = \sum_{i=0}^{4} a_i Z^i \tag{4.1.24}$$

From Eq. (4.1.6), one has

$$I_r \approx I_{exp} - I_{nr} \tag{4.1.25}$$

The coefficients a_i are obtained by fitting deviations $I_{exp} - I_{nr}$ to the fourth-order polynomial in Z, where I_{exp} is measured experimentally and I_{nr} is calculated from Eq. (4.1.19).

In the WBEPM Theory, after properly considering correlation and relativistic corrections, the ionization energy (of ground state and excited state) for an iso-spectrum-level series of an atomic system can be calculated using the following formula:

$$I_{cal} = \frac{R}{n'^2}[(Z-\sigma)^2 + g(Z-Z_0)] + \sum_{i=0}^{4} a_i Z^i \qquad (4.1.26)$$

4.1.3 Calculation of Ionization Energy

The ionization energies of the ground state and excited state for elements have been calculated using Eq. (4.1.26) in Refs. [6–9]. Partial results are listed here.

The calculated successive ionization energies and their comparisons with experimental data are given in Table 4.12 for elements with $Z = 2 - 18$. Among 171 pairs of compared data, 167 pairs have relative error equal to or lower than one thousandth. Only 4 pairs have relative error higher than one percent. However, with respect to absolute error, among 171 pairs of compared data, 68% have error equal to or smaller than 0.01 eV, 32% have error larger than 0.1 eV, and the maximum error is -0.64522 eV. The comparisons of the calculated values with experimental data for Be I iso-spectrum-level series, B I iso-spectrum-level series and C I iso-spectrum-level series are given in Tables 4.13, 4.14, and 4.15, respectively. Take B I series for example, among 81 calculated data, the maximum absolute error between calculation and experimental data is 0.133 eV, most absolute errors are all equal to or lower than 0.01 eV. However, the relative errors are all equal to or lower than one thousandth. It is roughly the same for Be I series and C I series. The values for I_{nr} and I_{cal} are given in Table 4.16 for C I iso-spectrum-level series. By comparing with experimental data, one can evaluate the error between calculations with higher-order relativistic corrections neglected and calculations with relativistic corrections included. Tables 4.17 and 4.18 list the calculated values using WBEPM Theory, calculated values using other theoretical methods and their comparisons with experimental data. The data in these tables show that WBEPM Theory gives more accurate results than other theoretical methods.

The following three points are particularly worth to mention:

(1) The calculated I_{nr} from Eq. (4.1.19) is not purely non-relativistic ionization energy, as the result of Eq. (4.1.12) comes from the non-relativity of the system and the central field approximation. However, experimental data are used when determining the parameter n', σ and g, so it includes electron correlation effect and partial relativistic effect with Z^2 as the highest power. The eventually obtained formula for I_{cal}, i.e. Eq. (4.1.26) by fitting $I_{exp} - I_{nr}$ also includes relativistic corrections with Z^4 as the highest power. Thus, it is obvious that the

4.1 Ionization Energy

Table 4.12 The comparison of calculated ionization energies with experimental data for the ground state of element with $Z = 2 - 18$[①]

Z	I_{cal} (I_1)	$I_{exp} - I_{cal}$	I_{cal} (I_2)	$I_{exp} - I_{cal}$	I_{cal} (I_3)	$I_{exp} - I_{cal}$	I_{cal} (I_4)	$I_{exp} - I_{cal}$	I_{cal} (I_5)	$I_{exp} - I_{cal}$	I_{cal} (I_6)	$I_{exp} - I_{cal}$
2	24.654 80	−0.067 39										
3	5.305 74	0.085 98	75.635 57	0.004 61								
4	9.232 22	0.090 48	18.235 22	−0.024 06	153.857 60	0.039 01						
5	8.229 85	−0.068 18	25.172 54	−0.017 70	37.980 61	−0.049 97	259.325 02	0.050 19				
6	11.160 62	−0.099 68	24.435 12	0.051 80	47.938 82	−0.051 02	64.544 37	−0.050 47	392.046 91	0.040 09		
7	14.280 80	0.253 34	29.632 92	0.031 62	47.504 63	0.055 39	77.531 40	−0.057 90	97.930 58	−0.040 38	552.037 35	0.034 45
8	13.406 87	0.211 19	35.182 43	−0.065 13	54.983 41	0.047 91	77.438 63	0.025 10	113.952 33	−0.053 33	138.144 95	−0.025 25
9	17.281 10	0.141 72	35.065 90	−0.095 08	62.874 92	−0.166 52	87.201 09	0.061 29	114.238 83	−0.003 97	157.205 43	−0.040 33
10	21.487 92	0.076 68	41.070 45	−0.107 17	63.616 66	−0.166 66	97.359 15	−0.239 15	126.277 86	0.067 86	157.908 46	−0.021 54
11	5.062 59	0.076 49	47.314 55	−0.028 15	71.768 81	−0.148 81	99.045 18	−0.135 18	138.638 21	−0.238 21	172.208 49	0.028 49
12	7.486 06	0.160 18	15.133 41	−0.098 13	80.177 13	−0.033 43	109.357 84	−0.092 34	141.340 76	−0.070 76	186.717 44	0.042 56
13	5.911 45	0.074 32	18.955 37	−0.126 81	28.507 28	−0.059 63	120.018 99	−0.026 99	153.823 12	0.001 88	190.495 93	−0.005 93
14	8.011 60	0.140 09	16.466 32	−0.120 47	33.642 24	−0.149 22	45.135 87	0.005 94	166.792 91	−0.025 91	205.154 15	0.115 85
15	10.380 18	0.106 51	19.953 10	−0.183 70	30.266 26	−0.063 56	51.528 41	−0.084 51	64.976 93	0.048 17	220.461 15	−0.040 15
16	10.322 96	0.037 05	23.433 45	− 0.095 55	35.038 28	−0.248 28	47.289 87	−0.067 87	72.597 82	−0.003 32	87.994 23	0.058 77
17	13.976 89	−0.009 25	23.871 96	−0.057 96	39.869 05	−0.259 05	53.290 66	0.174 64	67.518 57	0.281 43	96.836 64	0.193 36
18	15.777 27	−0.017 65	27.661 81	−0.031 81	40.762 08	−0.022 08	59.606 11	0.203 89	74.727 63	0.292 37	90.936 55	0.072 45

(continued)

Table 4.12 (continued)

Z	I_{cal}	$I_{exp} - I_{cal}$	I_{cal}	$I_{exp} - I_{cal}$	I_{cal}	$I_{exp} - I_{cal}$
Z	I_7		I_8		I_9	
8	739.315 43	−0.025 43				
9	185.194 82	−0.008 82	953.905 18	0.006 02		
10	207.296 21	−0.020 31	239.089 14	0.009 76	1 195.835 6	−0.007 06
11	208.452 20	0.047 80	264.231 95	0.018 05	299.838 49	0.025 51
12	224.990 64	−0.029 36	265.876 26	0.083 74	328.021 65	0.038 35
13	241.604 38	−0.115 62	284.624 87	−0.035 13	330.188 30	−0.058 30
14	246.506 43	−0.006 43	303.308 82	−0.231 18	351.114 61	−0.005 39
15	263.344 38	0.225 62	309.371 28	0.228 72	371.842 74	−0.287 26
16	280.995 42	−0.047 42	328.391 15	0.358 85	379.092 70	0.457 30
17	114.157 63	0.038 17	348.376 92	−0.096 92	400.295 76	−0.235 76
18	124.233 27	0.089 73	143.443 04	0.016 96	422.596 28	−0.146 28
	I_{10}		I_{11}		I_{12}	
	1 465.140 8	−0.019 87				
	367.455 06	0.044 94	1 761.859 8	−0.054 84		
	398.676 03	0.073 97	441.952 70	0.047 30	2 086.036 5	−0.056 55
	401.397 48	−0.027 48	476.207 56	0.152 44	523.346 84	0.073 16
	424.466 19	0.066 19	479.514 46	−0.054 46	560.630 45	0.169 55
	447.220 89	−0.279 61	504.688 82	−0.111 18	564.551 37	−0.111 37
	455.676 18	−0.046 18	529.458 20	0.178 20	591.794 59	−0.195 41
	479.063 41	−0.373 41	539.130 42	−0.170 42	618.574 86	0.314 86
Z	I_{13}		I_{14}		I_{15}	
14	2 437.719 9	−0.089 99				
15	611.654 56	0.085 44	2 816.964 1	−0.054 12		
16	651.960 62	0.239 38	706.894 56	0.115 44	3 223.827 8	−0.047 88
17	656.521 83	0.188 17	750.215 75	−0.455 75	809.087 16	0.312 84
18	685.798 49	−0.301 51	755.440 97	0.299 03	855.415 22	−0.645 22
	I_{16}		I_{17}		I_{18}	
	3 658.375 2	0.145 78				
	918.254 31	−0.224 31	4 120.675 0	0.210 65		

Data source Zheng NW, Zhou T, Wang T, et al. Phys Rev A, 2002, 65:052510 (Table II)
Notes [1] I_{cal} are calculated values using WBEPM Theory
I_{exp} are experimental data taken from Lide DR. CRC handbook of chemistry and physics, 81st edn. CRC Press, Inc., Florida, 2001

4.1 Ionization Energy

Table 4.13 The comparison of calculated ionization energies with experimental data for Be I iso-spectrum-level series of several excited-states[①]

Z	[He]2s3s $^1S_0$		[He]2s3p $^1P_1^0$		[He]2s3d $^1D_2$		[He]2s4d $^1D_2$	
	I_{cal}	I_{exp}	I_{cal}	I_{exp}	I_{cal}	I_{exp}	I_{cal}	I_{exp}
4	2.513	2.544	1.843	1.860	1.339	1.335	0.796	0.795
5	8.372	8.343	7.309	7.288	5.968	5.976	3.391	3.390
6	17.288	17.243	15.799	15.784	13.609	13.608	7.683	7.691
7	29.258	29.262	27.319	27.319	24.267	24.265	13.675	13.665
8	44.282	44.310	41.871	41.886	37.944	37.944	21.373	21.373
9	62.357	62.391	59.458	59.483	54.643	54.648	30.778	30.778
10	83.479	83.509	80.084	80.106	74.365	74.377	41.894	41.891
11	107.647	107.683	103.748	103.784	97.113	97.149	54.720	54.718
12	134.858	134.724	130.451	130.320	122.887	122.759	69.257	69.036
13	165.108	165.107	160.191	160.219	151.691	151.727	85.501	85.368
14	198.396	198.411	192.968	193.051	183.524	183.597	103.452	103.444
15	234.717	234.764	228.778	228.776	218.387	218.398	123.104	122.968
16	274.069		267.618	267.620	256.280	256.238	144.452	144.442
17	316.448	316.433	309.482	308.737	297.204	296.353	167.491	
18	361.850		354.366		341.159		192.211	
19	410.273		402.262	402.465	388.143	388.195	218.606	
20	461.711		453.164	453.410	438.155	438.160	246.664	
21	516.162		507.061	506.848	491.196		276.375	
22	573.621		563.946		547.262	547.279	307.726	
23	634.083		623.806	623.145	606.353		340.705	341.316

Data source Zhang NW, Wang T. Int J Quantum Chem, 2003, 93:344–350 (Table III)
Notes [①] I_{cal} are calculated values using WBEPM Theory
I_{exp} are experimental data taken from Fuhr JR, Martin WC, Musgrove A, et al. NIST Atomic Spectroscopic Database, Version 2.0, 1996; http://physics.nist.gov

results of calculations with Eq. (4.1.26) are relatively accurate. The results will be better if fitting is done in a smaller interval of Z.

(2) The unknown ionization energies for the members of a given iso-spectrum-level series can be predicted by Eq. (4.1.26).

(3) Combining Eqs. (4.1.1), (4.1.2), and (4.1.26) leads to

$$I_{cal}(Z) = \frac{R}{n'^2}[(Z-\sigma)^2 + g(Z-Z_0)] + \sum_{i=0}^{4} a_i Z^i$$
$$= T_{limit}(Z) - T(Z,n) \qquad (4.1.27)$$

Table 4.14 The comparison of calculated ionization energies with experimental data for B I isospectrum-level series of several excited-states[①]

Z		$1s^22s^22p$ $^2P^0_{3/2}$	$1s^22s^23s$ $^2S_{1/2}$	$1s^22s^23p$ $^2P^0_{3/2}$	$1s^22s^23d$ $^2D_{5/2}$	$1s^22s^24d$ $^2D_{5/2}$	$1s^22s^25d$ $^2D_{5/2}$	$1s^22s^26d$ $^2D_{5/2}$
5	I_{exp}	8.296	3.334	2.271	1.508	0.860	0.551	0.382
	I_{cal}	8.276	3.306	2.271	1.486	0.855	0.552	0.386
	$I_{exp} - I_{cal}$	0.020	0.028	0.000	0.022	0.005	−0.001	−0.004
6	I_{exp}	24.375	9.934	8.050	6.337	3.539	2.253	1.561
	I_{cal}	24.404	10.017	8.049	6.363	3.562	2.249	1.553
	$I_{exp} - I_{cal}$	−0.029	−0.083	0.001	−0.026	−0.023	0.004	0.008
7	I_{exp}	47.428	20.011	16.986	14.315	8.050	5.053	3.493
	I_{cal}	47.440	19.957	16.988	14.334	7.995	5.057	3.495
	$I_{exp} - I_{cal}$	−0.012	0.054	−0.002	−0.019	0.055	−0.004	−0.002
8	I_{exp}	77.366	33.075	29.029	25.396	14.110	8.970	6.201
	I_{cal}	77.358	33.037	29.027	25.389	14.145	8.969	6.204
	$I_{exp} - I_{cal}$	0.008	0.038	0.002	0.007	−0.035	0.001	−0.003
9	I_{exp}	114.151	49.182	44.124	39.540	21.994	13.980	9.662
	I_{cal}	114.139	49.191	44.125	39.519	22.007	13.982	9.676
	$I_{exp} - I_{cal}$	0.012	−0.009	−0.001	0.021	−0.013	−0.002	−0.014
10	I_{exp}	157.769	68.339	62.270	56.710	31.585	20.094	13.914
	I_{cal}	157.765	68.376	62.270	56.713	31.578	20.091	13.907
	$I_{exp} - I_{cal}$	0.004	−0.037	0.000	−0.003	0.007	0.003	0.007
11	I_{exp}	208.239	90.552	83.476	76.994	42.882	27.290	18.875
	I_{cal}	208.223	90.574	83.470	76.961	42.859	27.292	18.895
	$I_{exp} - I_{cal}$	0.016	−0.022	0.006	0.033	0.023	−0.002	−0.020
12	I_{exp}	265.549	115.852		100.312	55.950	35.544	24.678
	I_{cal}	265.505	115.786	107.758	100.254	55.853	35.581	24.636
	$I_{exp} - I_{cal}$	0.044	0.066		0.058	0.097	−0.037	0.042
13	I_{exp}	329.520	144.024		126.497	70.443	44.820	31.073
	I_{cal}	329.603	144.042	135.192	126.582	70.567	44.953	31.129
	$I_{exp} - I_{cal}$	−0.083	−0.018		−0.085	−0.124	−0.133	−0.056
14	I_{exp}	400.508			155.919	87.036	55.456	38.383
	I_{cal}	400.516	175.389	165.852	155.935	87.009	55.406	38.374
	$I_{exp} - I_{cal}$	−0.008			−0.016	0.027	0.050	0.009
15	I_{exp}	478.257	209.910		188.331	105.263	67.038	
	I_{cal}	478.245	209.902	199.843	188.304	105.191	66.935	46.372
	$I_{exp} - I_{cal}$	0.012	0.008		0.027	0.072	0.103	

(continued)

4.1 Ionization Energy

Table 4.14 (continued)

Z		$1s^22s^22p$ $^2P^0_{3/2}$	$1s^22s^23s$ $^2S_{1/2}$	$1s^22s^23p$ $^2P^0_{3/2}$	$1s^22s^23d$ $^2D_{5/2}$	$1s^22s^24d$ $^2D_{5/2}$	$1s^22s^25d$ $^2D_{5/2}$	$1s^22s^26d$ $^2D_{5/2}$
16	I_{exp}	562.810			223.717	125.125	79.536	55.111
	I_{cal}	562.796	247.678	237.295	223.681	125.128	79.536	55.124
	$I_{exp} - I_{cal}$	0.014			0.036	−0.003	0.000	−0.013
17	I_{exp}	654.189			262.078			
	I_{cal}	654.176	288.834	278.360	262.055	146.838	93.207	64.633
	$I_{exp} - I_{cal}$	0.013			0.023			
18	I_{exp}	752.391						
	I_{cal}	752.400	333.515	323.213	303.419	170.341	107.944	74.902
	$I_{exp} - I_{cal}$	−0.009						
19	I_{exp}	857.476			347.776			
	I_{cal}	857.481	381.886	372.057	347.763	195.660	123.743	85.935
	$I_{exp} - I_{cal}$	−0.005			0.013			

Data source Zhang NW, Wang T. Int J Quantum Chem, 2004, 98:495–501 (Table II)
Notes [1] I_{cal} are calculated values using WBEPM Theory
I_{exp} are experimental data taken from Fuhr JR, Martin WC, Musgrove A, et al. NIST Atomic Spectroscopic Database, Version 2.0, 1996; http://physics.nist.gov

4.1.4 The Successive Ionization Energies of the $4f^n$ Electrons for the Lanthanides [10]

Faktor and Hanks first calculated the third ionization energies for the lanthanides using a Born-Haber cycle and the thermodynamic data [33]. Then Sugar and Reader derived the third and fourth ionization energies [34]. Sugar further obtained the fifth ionization energy [35]. Sinha proposed the "inclined W" theory and two simple correlations, i.e.

$$IP\left(\sum IP\right) = W_1 L + k_1 \qquad (4.1.28)$$

and

$$IP\left(\sum IP\right) = mL + C \qquad (4.1.29)$$

The former is called the series correlation and the latter is called the successive correlation, and using these correlations he obtained all predicted values for the sixth to seventeenth ionization energies for the lanthanides [36–38]. With data in Refs. [25, 26], we calculated the successive ionization energies of the $4f^n$ electrons for the lanthanides based on the finite differential law and Eq. (4.1.19). All the calculated

Table 4.15 The comparison of calculated ionization energies with experimental data for C I iso-spectrum-level series of several excited-states[①]

Z	$1s^22s^22p^2\,^3P_2$		$1s^22s^22p3s\,^3P_2^0$		$1s^22s^22p3d\,^3D_3^0$		$1s^22s^22p3d\,^1F_3^0$		$1s^22s^22p4d\,^3D_3^0$	
	I_{cal}	I_{exp}	I_{cal}	I_{exp}	I_{cal}	I_{exp}	I_{cal}	I_{exp}	I_{cal}	I_{exp}
6	11.243	11.26	3.740	3.773	1.529	1.550	1.537	1.524	0.870	0.864
7	29.615	29.59	11.180	11.118	6.375	6.355	6.125	6.127	3.514	3.527
8	54.896	54.898	21.748	21.753	14.354	14.349	13.773	13.795	7.906	7.901
9	87.060	87.06	35.411	35.421	25.436	25.429	24.470	24.476	14.036	14.063
10	126.084	126.08	52.139	52.133	39.596	39.584	38.205	38.193	21.889	21.819
11	171.944	172.0	71.908	71.905	56.809	56.815	54.968	54.962	31.450	31.492
12	224.623	224.7	94.694	94.725	77.051	77.100	74.745	74.774	42.696	42.711
13	284.103	284.1	120.482	120.493	100.300	100.329	97.523	97.533	55.603	55.598
14	350.372	350.3	149.256	149.215	126.534	126.535	123.288	123.268	70.141	70.132
15	423.416	423.3	181.009	180.955	155.735	155.687	152.024	151.918	86.280	
16	503.228	503.241	215.733	215.728	187.883	187.944	183.714	183.740	103.981	103.986
17	589.801		253.427		222.961		218.341		123.207	
18	683.131	683.17	294.093	294.091	260.953	260.987	255.886	256.028	143.912	
19	783.217	783.18	337.737		301.846	301.691	296.330	296.248	166.050	
20	890.060	890.09	384.369		345.625	345.743	339.653	339.717	189.570	
21	1 003.665	1 003.6	434.004		392.279	392.410	385.833	385.839	214.416	
22	1 124.037	1 123.8	486.658	486.886	441.797	441.756	434.848	434.689	240.530	
23	1 251.185	1 251.2	542.354	542.307	494.169	494.201	486.674	486.762	267.850	

Data source Zhang NW, Wang T. Chem Phys Lett, 2003, 376:557–565 (Table 2)

Notes ① I_{cal} are calculated values using WBEPM Theory

I_{exp} are experimental data taken from Fuhr JR, Martin WC, Musgrove A, et al. NIST Atomic Spectroscopic Database, Version 2.0, 1996; http://physics.nist.gov

4.1 Ionization Energy

Table 4.16 The comparison of calculated I_{nr}, I_{cal} with I_{exp} for C I iso-spectrum-level series of several excited states[①]

Z	I_{exp}	I_{nr}	$I_{exp} - I_{nr}$	I_{cal}	$I_{exp} - I_{cal}$
	$1s^22s^22p^2\,{}^3P_2$				
6	11.26	11.260	0	11.243	0.017
7	29.59	29.709	−0.119	29.615	−0.025
8	54.898	54.990	−0.092	54.896	0.002
9	87.06	87.099	−0.039	87.060	0
10	126.08	126.035	0.045	126.084	−0.004
11	172.0	171.799	0.201	171.944	0.056
12	224.7	224.390	0.310	224.623	0.077
13	284.1	283.809	0.291	284.103	−0.003
14	350.3	350.054	0.246	350.372	−0.072
15	423.3	423.128	0.172	423.416	−0.116
16	503.241	503.028	0.213	503.228	0.013
17	–	589.757	–	589.801	–
18	683.17	683.312	−0.142	683.131	0.039
19	783.18	783.695	−0.515	783.217	−0.037
20	890.09	890.905	−0.815	890.060	0.03
21	1 003.6	1 004.493	−1.343	1 003.665	−0.065
22	1 123.8	1 125.808	−2.008	1 124.037	−0.237
23	1 251.2	1 253.501	−2.301	1 251.185	0.015
Z	$1s^22s^22p3s\,{}^3P_2^0$				
6	3.773	3.773	0	3.740	0.033
7	11.118	11.300	−0.182	11.180	−0.062
8	21.753	21.859	−0.106	21.748	0.005
9	35.421	35.448	−0.027	35.411	0.01
10	52.133	52.068	0.065	52.139	−0.006
11	71.905	71.718	0.187	71.908	−0.003
12	94.725	94.399	0.326	94.694	0.031
13	120.493	120.111	0.382	120.482	0.011
14	149.215	148.854	0.361	149.256	−0.041
15	180.955	180.627	0.328	181.009	−0.054
16	215.728	215.431	0.297	215.733	−0.005
Z	$1s^22s^22p3s\,{}^3P_2^0$				
17	–	253.266	–	253.427	–
18	294.091	294.132	−0.041	294.093	−0.002
19	–	338.028	–	337.737	–

(continued)

Table 4.16 (continued)

Z	I_{exp}	I_{nr}	$I_{exp} - I_{nr}$	I_{cal}	$I_{exp} - I_{cal}$
20	–	384.954	–	384.369	–
21	–	434.912	–	434.004	–
22	486.886	487.900	−1.014	486.658	0.028
23	542.307	543.919	−1.612	542.354	−0.047
Z	$1s^22s^22p3d^1F_3^0$				
6	1.524	1.524	0	1.537	−0.013
7	6.127	6.175	−0.048	6.125	0.002
8	13.795	13.839	−0.044	13.773	0.022
9	24.476	24.517	−0.041	24.470	0.006
10	38.193	38.209	−0.016	38.205	−0.012
11	54.962	54.915	0.047	54.968	−0.006
12	74.774	74.635	0.139	74.745	0.029
13	97.533	97.369	0.164	97.523	0.01
14	123.268	123.177	0.151	123.288	−0.02
15	151.918	151.878	0.04	152.024	−0.106
16	183.740	183.653	0.087	183.714	0.026
17	–	218.442	–	218.341	–
18	256.028	256.245	−0.217	255.886	0.142
19	296.248	297.062	−0.814	296.330	−0.082
20	339.717	340.892	−1.175	339.653	0.064
21	385.839	387.737	−1.898	385.833	0.006
22	434.089	437.595	−2.906	434.848	−0.159
23	486.762	490.467	−3.705	486.674	0.088

Data source Zhang NW, Wang T. Chem Phys Lett, 2003, 376:557–565 (Tables 2 and 3)
Notes [①] I_{nr} are calculated non-relativistic ionization energies
I_{cal} are calculated ionization energies with relativistic corrections included
I_{exp} are experimental data taken from Fuhr JR, Martin WC, Musgrove A, et al. NIST Atomic Spectroscopic Database, Version 2.0, 1996; http://physics.nist.gov

results are listed in Table 4.19. The plot of the calculated values I against the total orbital angular momentum L of the lanthanide ions ($4f^m$) also shows an "inclined W" distribution as proposed by Sinha (see Fig. 4.2).

4.1 Ionization Energy

Table 4.17 The comparison of calculated ground-state ionization energies using WBEPM theory and other theoretical methods with experimental data for C, N, O and F atoms

Element	Theory	Calculations			
		I_1	I_2	I_3	I_4
C	HF[1]	10.81	24.16	45.80	64.29
	G2[1]	11.18	24.22	47.93	64.29
	B3LYP[1]	11.54	25.04	47.25	65.06
	BLYP[1]	11.41	24.93	47.12	65.04
	SLYP[1]	11.20	24.23	46.21	64.02
	SPW91[1]	11.28	24.20	46.15	63.75
	WBEPM theory[2]	11.161	24.435	47.939	64.544
	Exp. value[1]	11.24	24.39	47.86	64.49[3]
N	HF[1]	13.91	29.19	47.32	75.03
	G2[1]	14.48	29.46	47.27	77.49
	B3LYP[1]	14.67	29.99	48.34	76.60
	BLYP[1]	14.51	29.83	48.24	76.48
	SLYP[1]	14.55	29.41	47.30	75.32
	SPW91[1]	14.78	29.52	47.21	75.29
	WBEPM theory[2]	14.281	29.633	47.505	77.531
	Exp. value[1]	14.54	29.59	47.43	77.45
O	HF[1]	12.05	34.55	54.61	77.47
	G2[1]	13.53	35.01	54.77	77.30
	B3LYP[1]	14.16	35.29	55.47	78.60
	BLYP[1]	14.17	35.06	55.29	78.52
	SLYP[1]	13.47	35.03	54.72	77.34
	SPW91[1]	13.37	35.28	54.77	77.19
	WBEPM theory[2]	13.407	35.182	54.983	77.439
	Exp. value[1]	13.618[3]	35.11	54.90	77.414[3]
F	HF[1]	15.70	33.37	62.23	87.00
	G2[1]	17.39	34.77	62.59	87.06
	B3LYP[1]	17.76	35.56	62.93	87.90
	BLYP[1]	17.73	35.53	62.69	87.71
	SLYP[1]	17.29	34.58	62.56	86.99
	SPW91[1]	17.40	34.49	62.80	87.02
	WBEPM theory[2]	17.281	35.066	62.875	87.201
	Exp. value[1]	17.45	34.98	62.65	87.16

Notes [1] Jursic BS. Int J Quantum Chem, 1997, 64:255 (Tables II–V)
[2] Zheng NW, Zhou T, Wang T, et al. Phys Rev A, 2002, 65:052510 (Table II)
[3] The experimental data from the paper of Jursic may have error. These three data are taken from http://physics.nist.gov (CIV is 520178.4 cm^{-1}, OI is 109837.02 cm^{-1}, and OIV is 624382.0 cm^{-1})

Table 4.18 The comparison of calculated ionization energies using WBEPM theory and other theoretical methods with experimental data for Be I [He]2s2p $^3P_1^0$ iso-spectrum-level series[①]

Z	Experimental results	WBEPM theory	RCI method (CIV3)	MCDF method	MBPT method
4	6.598	6.571			6.767
5	20.525	20.558	20.519		20.613
6	41.392	41.415	41.374		41.448
7	69.133	69.134	69.096	69.129	69.171
8	103.723	103.707	103.681	103.719	103.750
9	145.155	145.129	145.104	145.149	145.173
10	193.426	193.399	193.366	193.423	193.441
11	248.550	248.515	248.478	248.546	248.561
12	310.341	310.480	310.268	310.337	310.349
13	379.305	379.298	379.228	379.301	379.311
14	455.017	454.975	454.941	455.013	455.022
15	537.510	537.519		537.506	537.513
16	626.930	626.940	626.860	626.927	626.934
17	722.498	723.251		722.528	722.545
18	826.508	826.465	826.456	826.505	826.512
19	937.008	936.600		937.003	937.011
20	1 053.933	1 053.675	1 053.915	1 053.918	1 053.928
21	1 177.000	1 177.710		1 176.990	1 177.002
22	1 308.703	1 308.728		1 308.695	1 308.710
23	1 446.574	1 446.753		1 446.553	1 446.571

Data source Zhang NW, Wang T. Int J Quantum Chem, 2003, 93:344–350 (Table II)
Notes [①] The experimental data are taken from Fuhr JR, Martin WC, Musgrove A, et al. NIST Atomic Spectroscopic Database, Version 2.0, 1996; http://physics.nist.gov
The data for RCI, MCDF, and MBPT method are obtained by subtracting the energies of the corresponding excited-state energy levels using these methods from the ground-state experimental ionization energies. The values for RCI method are taken from Kingston AE, Hibbert AJ. Phys B, 2000, 33:693; the values for MCDF method are taken from JÖnsson P, Fischer CF, Träbert E. J Phys B, 1998, 31:3497; the values for MBPT method are taken from Safronova MS, Johnson WR, Safronova UI. Phys Rev A, 1996, 53:4036

4.2 Energy Level [39–50]

4.2.1 Introduction

It is well known that metal and alloy are collections of ordered atoms, and metal ions can form coordination compounds or coordination polymers of 0-dimensional (D), one-D, two-D, three-D structures by linking with inorganic or organic ligands.

4.2 Energy Level

Table 4.19 The calculated successive ionization energies of the $4f^n$ electrons for the lanthanides (eV)

Z	Element	I_3	I_4	I_5	I_6	I_7	I_8	I_9
57	La							
58	Ce	–	36.69					
59	Pr	21.60	38.91	57.54				
60	Nd	22.13	40.39	60.00	80.92			
61	Pm	22.34	41.11	61.73	83.63	106.85		
62	Sm	23.42	41.44	62.64	85.62	109.80	135.33	
63	Eu	24.75	42.73	63.08	86.71	112.04	138.52	166.35
64	Gd	–	44.18	64.59	87.27	113.33	141.02	169.78
65	Tb	21.91	39.76	66.15	89.00	114.01	142.50	172.54
66	Dy	22.77	41.56	62.08	90.67	115.95	143.28	174.21
67	Ho	22.88	42.57	63.75	86.94	117.74	145.44	175.11
68	Er	22.77	42.76	64.91	88.49	114.35	147.35	177.48
69	Tm	23.64	42.80	65.20	89.80	115.77	144.30	179.50
70	Yb	25.08	43.84	65.37	90.17	117.23	145.59	176.80
71	Lu	–	45.19	66.59	90.47	117.70	147.21	177.97

Z	Element	I_{10}	I_{11}	I_{12}	I_{13}	I_{14}	I_{15}	I_{16}	I_{17}
64	Gd	199.92							
65	Tb	203.59	236.03						
66	Dy	206.60	239.94	274.69					
67	Ho	208.46	243.21	278.84	315.89				
68	Er	209.48	245.26	282.36	320.29	359.64			
69	Tm	212.07	246.39	284.60	324.06	364.27	405.93		
70	Yb	214.20	249.20	285.85	326.49	368.30	410.81	454.76	
71	Lu	211.84	251.44	288.87	327.85	370.93	415.09	459.88	506.14

Data source Zheng NW, Xin HW. J Phys B: At Mol Opt Phys, 1991, 24:1187

All species includes atoms (atomic ions), for instance, metal cations and inorganic anions form inorganic crystals by periodic arrangements in three-dimensional space, atoms are held together to form molecules by chemical bonds, etc. Their chemical, physical and spectroscopic properties, etc., are all closely related with atomic structures, and atomic energy level and transition between energy levels are the main contents of atomic structure. Thus, the studies of atomic energy level are essential to chemistry, astrophysics, physics, material science, etc. In addition, high technology, space technology, and military technology such as laser, plasma, nuclear fusion, isotope separation, etc., also put massive demands on all kinds of atomic or molecular data (including energy level, lifetime, section, etc.). Therefore, the studies and calculations of atomic energy level structure are very necessary to the field of technology.

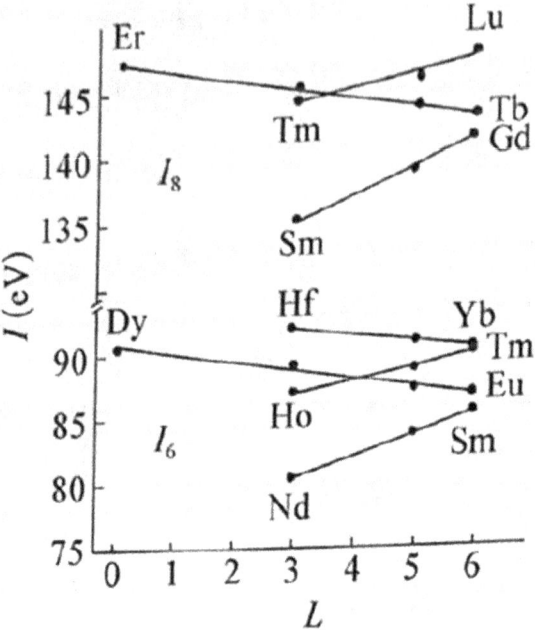

Fig. 4.2 The plot of the calculated ionization energies I against the total orbital angular momentum L of the lanthanide ions shows the profile of an inclined W. (Reproduced from Zheng NW, Xin HW (1991) J Phys B: At Mol Opt Phys 24:1187)

Generally atomic energy level structure is very complicated. One electron has a ground-state electronic structure (i.e. ground-state electron configuration), and when one or more electrons are excited to different excited-state energy levels, different excited-state electronic structures (i.e. excited-state electron configuration) are formed. One electron configuration can be divided into several spectral terms, and one spectral term can again split into several energy levels, and energy level will further split with existence of an external magnetic field. The theory of multiplicity of atomic structure has explained these complicated phenomena very well. However, it is still difficult to precisely calculate spectrum levels of an atom truly from theory, especially the complicated atomic energy levels, energy levels of high Rydberg states and atomic energy levels of higher ionization states. At present, the main theoretical methods for calculating atomic energy levels are Density Functional Theory (DFT), Many-Body Perturbation Theory (MBPT), Configuration Interaction (CI) method, Multichannel Quantum Defect Theory (MQDT), the Multiconfiguration Hartree–Fock (MCHF) method, etc. [51–71]. Many works have been published, but due to the complexity in the calculations, difficulties still exist in the field of calculating complicated atomic energy levels, energy levels of high Rydberg states and atomic energy levels of higher ionization states, and many gaps remain.

With respect to the calculations of atomic energy levels, we proposed formulae for calculating energy levels in WBEPM Theory, and many calculations have been done based on those formulae. In the last two parts of this section, we will describe the formulae for calculating energy levels and their applications, respectively.

4.2.2 *Formulae for Calculating Energy Levels*

Before introducing our formulae for calculating energy levels, first we want to describe the following four considerations.

Firstly, in Chap. 2, we have claimed that for an N-electron system, the WBE theory just renames every electron of the system the name of the weakest bound electron. Thus, the system remains at the same status, and any observable physical quantity in the system, especially Hamiltonian, also remains the same.

Specifically for the problem of atomic energy levels that will be discussed in this section, the electronic configurations, coupling schemes, term symbols of energy levels, etc., all keep the same. Take, for example, an excited atom with electronic configuration C I [He]$2s^2$2p5d, generally speaking the system includes two 1s electrons, two 2s electrons, one 2p electron and one excited 5d electron. While from the WBE theory point of view, 5d electron has the weakest association with the system comparing with other electrons, so it is the weakest bound electron of the present system. Thus, this system includes two 1s electrons, two 2s electrons, one 2p electron and one weakest bound 5d electron. The two statements just use different names when discussing the behavior involving 5d electron and rename them.

Secondly, according to the WBE theory, whether for an atom with univalent or multivalent electrons or for an atomic system with one or two or many electrons excited, there is only one weakest bound electron in the current system. For a ground state such as the ground-state carbon atom, its electronic configuration is [He]$2s^2$$2p^2$. Two 2p electrons with higher energy are thought to be equivalent, but when a single electron is excited, the electron which is excited first is the weakest bound electron of the present system. It is the same for excite states, for instance, for the singly excited state with C[He]$2s^2$2p3s, 3s electron is the weakest bound electron of the present system, while for the doubly excite state with C[He]$2s^2$3s3d, both 3s and 5d electrons are excited electrons, but because 5d electron has higher energy and has weakest association with the system, 5d electron is the weakest bound electron in the present system. Thus, the WBE theory conceptually breaks the constraints of equivalent electrons and boundaries of single-electron, double-electron and multi-electron excitations in other theories, and treats the problems of univalent electron, multivalent electrons, single-electron excitation and double (multi)-electron excitation uniformly as the one-electron problem of the weakest bound electron.

Thirdly, the phenomenon of energy level perturbation due to configuration interaction is very common in the complicated structures of atomic energy levels. Configuration interaction includes: ① the perturbation of one spectral term in one configuration caused by a certain spectral term of another configuration; ② the atomic states generated by configurations with different principle quantum numbers in a term series may overlap and then interact with each other; ③ the interactions between the energy level in a series of discrete energy levels and the continuous energy level in another series, and these interactions will make the energy level move, i.e. the energy level is perturbed [72, 73]. Therefore, energy level perturbation should be shown in the given formula for the energy level calculation.

Fourthly, hydrogen spectrum and the spectra of alkali-metal atoms are relatively simple and similar to each other. While the structures of energy levels are very complicated for complicated atoms, in order to study the law of energy levels and accurately calculate them, a proper and definite method is required for energy level classification. In Ref. [74], Rydberg configuration series and Rydberg level series were defined. The definition of Rydberg configuration series is definite, but the definition of Rydberg level series is not clear. Thus, here we define a method to classify energy levels (maybe the classification method that we define is the same as Rydberg level series, but in this book we will use our classification method defined below).

Based on the above four considerations, let's first introduce the concept of "spectrum-level-like series". The so-called spectrum-level-like series is a series of energy levels with the same spectral level symbol in a given electronic configuration series of an atom. For example, for a LS coupling type carbon atom, C[He]$2s^2 2pnd\ ^3D_1^0$ is a spectrum-level-like series. The first part of the symbol "C" represents a given atom carbon, the second part [He]$2s^2 2pnd$ represents an electronic configuration series of the given atom carbon, and the third part $^3D_1^0$ is a spectral level symbol.

A spectral-level-like series is denoted by the symbols of the above three parts. Then, for a jj coupling type Neon atom, Ne[He]$2s^2 2p^5 ns\ [3/2]_1^0$ is also a spectral-level-like series. It is worthy of attention that a spectral-level-like series is always with respect to a given electronic configuration because different electronic configurations can have the same spectral level symbol. At the same time, from the above notations we can see that in a spectral-level-like series, the only difference between different energy levels is the different principle quantum number n, so the definition of spectral-level-like series can be used to classify complicated atomic energy levels.

Let's restate the general definition of ionization energy: "the energy required to completely remove one weakest bound electron from a certain energy level of a free particle (atom, molecule), i.e. the energy difference between series limit and energy level is called ionization energy." According to this definition, for a spectral-level-like series, one has

$$I_{exp} = T_{limit} - T(n) \quad (4.2.1)$$

T_{limit} represents the series limit of a given spectral-level-like series, $T(n)$ is the energy of one energy level in this series, and both T_{limit} and $T(n)$ are determined with respect to the ground-state energy level. I_{exp} represents the ionization energy of the weakest bound electron at the given energy level. Then

$$I_{exp} = -\varepsilon_\mu \quad (4.2.2)$$

ε_μ is the energy of the weakest bound electron at the given energy level.

$$\varepsilon_\mu = \varepsilon_\mu^0 + \Delta E_c + \Delta E_r \quad (4.2.3)$$

4.2 Energy Level

where ε_μ^0 represents the non-relativistic one-electron energy of the weakest bound electron in the central field approximation [refer to Eq. (2.4.9)]. Due to insufficient consideration of correlation effects and exclusion of relativistic effects, ΔE_r and ΔE_c must be added so that the sum will equal the true energy ε_μ of the weakest bound electron. ΔE_r and ΔE_c naturally represent electron correlation effects and relativistic corrections for energy, respectively. So

$$T(n) = T_{limit} - I_{exp}$$
$$= T_{limit} + \varepsilon_\mu^0 + \Delta E_c + \Delta E_r \qquad (4.2.4)$$

Under a given analytical potential (refer to Chap. 3)

$$\varepsilon = -\frac{RZ'^2}{n'^2} \approx \varepsilon_\mu^0 \qquad (4.2.5)$$

or

$$\varepsilon = -\frac{Z'^2}{2n'^2} \approx \varepsilon_\mu^0 (\text{a.u.})$$

Substituting it into Eq. (4.2.4), then

$$T(n) = T_{limit} - \frac{RZ'^2}{n'^2} + \Delta E_c + \Delta E_r \qquad (4.2.6)$$

Equation (4.2.6) is the formula for calculating atomic energy level within WBEPM theory. The energy of the atomic energy level can be calculated using Eq. (4.2.6).

Equations (4.2.4) and (4.2.6) are in agreement with Ritz's combination principle [74–76] and the description in Ref. [77].

4.2.3 Methods for Parameter Characterization

Related parameters need to be determined when calculating the energy of a given atomic energy level using Eq. (4.2.6). To maintain simplicity and accuracy, we proposed two different ways to determine these parameters.

4.2.3.1 Energy Level Calculation by the Formula of Ionization Energy

From Eq. (4.2.6), we get

$$\frac{RZ'^2}{n'^2} - \Delta E_c - \Delta E_r = T_{limit} - T(n) \qquad (4.2.7)$$

We have illustrated in Sect. 4.1 that after considering electron correlation and relativistic corrections, for an iso-spectrum-level series of an atomic system, the ionization energy of the weakest bound electron at the energy level can be represented by the following formula

$$I_{cal} = \frac{R}{n'^2}[(Z-\sigma)^2 + g(Z-Z_0)] + \sum_{i=0}^{4} a_i Z^i \quad (4.2.8)$$

For an iso-spectrum-level series, combining Eqs. (4.2.7) and (4.2.8) leads to

$$\frac{R}{n'^2}[(Z-\sigma)^2 + g(Z-Z_0)] + \sum_{i=0}^{4} a_i Z^i = T_{limit} - T(n) \quad (4.2.9)$$

When Z (i.e. one member of a given iso-spectrum-level series) is given, the ionization energy of the weakest bound electron can be calculated by using Eq. (4.2.8), and then after taking experimental data of T_{limit} and substituting it into Eq. (4.2.9), $T(n)$ can be obtained. For example, the results of calculation are given in Table 4.20 for some iso-spectrum-level series of beryllium-like atoms and ions. For more examples, readers can refer to Sect. 4.1.

Table 4.20 Atomic energy level calculated from ionization energy for some iso-spectrum-level series of beryllium-like atoms and ions (cm^{-1})

Z	I_{cal} (eV)[1]	T_{limit}[2]	$T_{cal}(n)$[1]	$T_{exp}(n)$[2]
Be I 2s3s $^1S_0$ series				
4	2.513	75 192.64	54 924.1	54 677.26
5	8.372	202 887.4	135 363.2	137 622.25[3]
6	17.288	386 241.0	246 805	247 170.26
7	29.258	624 866	388 886.2	388 854.6
8	44.282	918 657.0	561 501.5	561 276.4
9	62.357	1 267 606	764 666.9	764 392
10	83.479	1 671 750	998 451.9	998 183.1
11	107.647	2 131 300	1 263 075.3	1 262 780
12	134.858	2 644 700	1 557 005.6	1 558 080
13	165.108	3 216 100	1 884 424.8	1 884 420
14	198.396	3 842 100	2 241 941.2	2 241 810
15	234.717	4 523 000	2 629 894.9	2 629 500
16	274.069	5 260 000	3 049 502.2	
17	316.448	6 053 000	3 500 695.3	
Be I 2s3p $^1P_1^0$ series				
4	1.843	75 192.64	60 327.96	60 187.34
5	7.309	202 887.4	143 936.8	144 102.94

(continued)

4.2 Energy Level

Table 4.20 (continued)

Z	I_{cal} (eV)[1]	T_{limit}[2]	$T_{cal}(n)$[1]	$T_{exp}(n)$[2]
6	15.799	386 241.0	258 814.5	258 931.29
7	27.319	624 866	404 525.2	404 522.4
8	41.871	918 657.0	580 947.3	580 824.9
9	59.458	1 267 606	788 048.7	787 844
10	80.084	1 671 750	1 025 834.2	1 025 620.6
11	103.748	2 131 300	1 294 522.7	1 294 230
12	130.451	2 644 700	1 592 550.2	1 593 600
13	160.191	3 216 100	1 924 082.9	1 923 850
14	192.968	3 842 100	2 285 720.6	2 285 040
15	228.778	4 523 000	2 677 795.8	2 677 800
16	267.618	5 260 000	3 101 532.6	3 101 500
17	309.482	6 053 000	3 556 879.4	3 557 100
Be I 2s3d $^1D_2$ series				
4	1.339	75 192.64	64 392.96	64 428.31
5	5.968	202 887.4	154 752.6	154 686.12
6	13.609	386 241.0	276 477.9	276 482.86
7	24.267	642 866	429 141.0	429 159.6
8	37.944	918 657.0	612 620.5	612 615.6
9	54.643	1 267 606	826 884.0	826 843
10	74.365	1 671 650	1 071 960.7	1 071 914
11	97.113	2 131 300	1 348 037.1	1 347 740
12	122.887	2 644 700	1 653 557.5	1 654 580
13	151.691	3 216 100	1 992 639.4	1 992 340
14	183.524	3 842 100	2 361 891	2 361 290
15	218.387	4 523 000	2 761 704	2 761 500
16	256.280	5 260 000	3 192 979	3 193 300
17	297.204	6 053 000	3 655 907	3 656 700
Be I 2s4d $^1D_2$ series				
4	0.796	75 192.64	68 772.5	68 780.86
5	3.391	202 887.4	175 537.4	175 547.01
6	7.683	386 241.0	324 273.9	324 212.49
7	13.675	324 866	514 570.6	514 647.7
8	21.373	918 657.0	746 273.5	746 274.9
9	30.778	1 267 606	1 019 366.7	1 019 364
10	41.894	1 671 750	1 333 854.8	1 333 900

(continued)

Table 4.20 (continued)

Z	I_{cal} (eV)①	T_{limit}②	$T_{cal}(n)$①	$T_{exp}(n)$②
11	54.720	2 131 300	1 689 956.99	1 689 970
12	69.257	2 644 700	2 086 109	2 087 890
13	85.501	3 216 100	2 526 493	2 527 560
14	103.452	3 842 100	3 007 710	3 007 770
15	123.104	4 523 000	3 530 107	3 531 200
Be I 2s4 d $^1D_2$ series				
16	144.452	5 260 000	4 094 825	4 095 000
17	167.491	6 053 000	4 702 104.9	

Notes ① The data of I_{cal}(eV) were taken from Zheng NW, Wang T. Int J Quantum Chem, 2003, 93:344, which were obtained by calculations based on $I_{cal} = \frac{R}{n^2}[(Z-\sigma)^2 + g(Z-Z_0)] + \sum_{i=0}^{4} a_i Z^i$; $T_{cal}(n)$ was obtained by converting I_{cal}(eV) using 1 eV = 8065.479 cm^{-1}
② http://physcis.nist.gov (Select "Physical Reference Data"), Version 3.0, 2007
③ $T_{exp}(n)$ calculated from Fuhr JR, Martin WC, Musgrove A, et al. NIST Atomic Spectroscopic Database, Version 2.0, 1996; http://physics.nist.gov (Select "Physical Reference Data") is equal to 135 597.1 cm^{-1}. The value 137 622.25 cm^{-1} given in ③ only includes 81% of $^1S_0$ part

4.2.3.2 Energy Level Calculation by Introducing the Concept of Quantum Defect

Before describing this method, let's first introduce Martin's work.

In 1980, Martin proposed unperturbed series formulas [78] for the ns through ni series of Ritz type for the spectrum of atomic sodium (Na I):

$$E(nl) = \text{limit} - R_{Na}(n^*)^{-2} \quad (4.2.10)$$

where $E(nl)$ represents the energy of the nl level with respective to the ground level and the series limit is also determined with respect to the ground level. R_{Na} is the Rydberg constant for Na atom, and the factor that multiplies R_{Na} should be the atomic core charge Z_c ($Z_c = 1$). n^* is the effective principle quantum number.

$$n^* = n - \delta \quad (4.2.11)$$

where the quantum defect is

$$\delta = a + bm^{-2} + cm^{-4} + dm^{-6} \quad (4.2.12)$$

and

$$m = n - \delta_0 \quad (4.2.13)$$

4.2 Energy Level

For each series, δ_0 is the quantum defect of the lowest level for that series.

Martin did complete calculations for the spectrum of Na I using the series formulas and the precision is pretty high with wave-number uncertainties of a few percentages or less.

Due to conceptual limitations, the formulas are only used for a univalent-electron system. People think that a multivalent-electron system can't be treated like a univalent-electron system, because there are several equivalent (or identical) valence electrons in the multivalent-electron system. The idea that there is a boundary between the univalent-electron and the multivalent-electron system hinders seeking simple and accurate theoretical formulae for dealing with the atomic energy levels of univalent and multivalent electrons uniformly.

After carefully evaluating Martin's formulas, we combine Martin's formulas which have no consideration of energy level perturbations and Langer's formulas which consider perturbations, and then propose a three-step method with the concept of quantum defect introduced to calculate energy levels.

The following are the specific steps of this method:

Within WBEPM theory, for a spectral-level-like series, we have

$$T(n) = T_{limit} - \frac{RZ'^2}{n'^2} + \Delta E_c + \Delta E_r \qquad (4.2.14)$$

First neglecting ΔE_c and ΔE_r, then

$$T(n) = T_{limit} - \frac{RZ'^2}{n'^2} \qquad (4.2.15)$$

Step one, undergo transformation [1]

$$\frac{Z'}{n'} = \frac{Z_{net}}{n - \delta_n} \qquad (4.2.16)$$

then Eq. (4.2.15) becomes

$$T(n) = T_{limit} - \frac{RZ_{net}^2}{(n - \delta_n)^2} \qquad (4.2.17)$$

For a neutral atom, $Z_{net} = 1$. n is the principle quantum number and δ_n is the quantum defect.

Step two, combine Martin's formulas which is quite accurate for calculating unperturbed energy levels and Langer's formulae [76] which consider energy level perturbations, then we propose to take

$$\delta_n = \sum_{i=1}^{4} a_i m^{-2(i-1)} + \sum_{j}^{N} \frac{b_j}{m^{-2} - \varepsilon_j} \qquad (4.2.18)$$

where

$$m = n - \delta_0 \qquad (4.2.19)$$

δ_0 is the quantum defect of the lowest level in a given series.

$$\varepsilon_j = \frac{T'_{\text{limit}} - T_{j,\text{perturb}}}{RZ^2} \qquad (4.2.20)$$

$T_{j,\text{perturb}}$ is the energy of jth perturbed level and T'_{limit} is the series limit of the perturbed level series.

Step three, determine a_i and b_j by least square fitting.

In practical treatment, $a_i (i = 1 - 4)$ and $b_j (j = 1 - N)$ are determined by fitting the experimental data of the first six lowest energy levels in a given series plus additional N perturbed levels, i.e. $6 + N$ experimental data.

The series formulae and the three-step method to determine parameters have been applied in the following studies, such as analysis of the bound odd-parity spectrum of krypton [39], calculation of the energy levels of carbon group (C, Si, ge, Sn, Pb) [40], simple calculation of atomic energy levels of IB elements in the periodic table [41], study on the energy levels of atom neon [42], calculation of high Rydberg levels of atom Zn [43], study on the energy levels of atom gallium [44], calculation of the energy levels to high states in atom oxygen [45], theoretical study of energy levels and transition probabilities of Al II [46], theoretical analysis on autoionizing levels in Ca [47], and theoretical calculation of doubly excited levels of Sr I [48]. The comparison of massively calculated values with experimental data shows that the precision is pretty high with substantial deviations of fractional percentage or percentages of wave number (cm^{-1}) and the maximum deviation of more than 1 cm^{-1}, other than that many unknown energy level have been predicted.

Although we mentioned previously to neglect ΔE_c and ΔE_r, the parameters are obtained directly from or by fitting the experimental data, thus, we can think that these parameters have included considerable amount of electron correlation and relativistic effects. In other words, this method does not neglect ΔE_c and ΔE_r but largely include corrections for ΔE_c and ΔE_r. Thus, it is not hard to understand the high precision of the energies of the energy levels calculated by this method.

If a level series is not perturbed, the second term on the right side of Eq. (4.2.18) doesn't exist. Then Eqs. (4.2.17) and (4.2.18) return to the same form as the series formulae for atom Na proposed by Martin. Although the form is the same in this case, our formula is conceptually different from Martin's because it has broken the limitation of univalent-electron system and extended to multivalent-electron system. Zhang et al. have calculated the energy levels for Eu I, Au I, Pb I, B, Ru, Pb II and Ca II using unperturbed series formulas within WBEPM theory [79–85]. In the series that they calculated, most series (or systems) are unperturbed level series. The calculated results have maximum deviation of more than 1 cm^{-1}. Some series (or systems) are originally perturbed level series, but are still calculated by using unperturbed series formulas, thus, the maximum deviation is increased by two magnitudes, i.e. hundreds of wave number. If Eq. (4.2.18) is used for perturbation correction, the deviation will be less than 1 cm^{-1}.

Although the concept of quantum defect is introduced into the second method, readers can completely understand that this is not identical to the quantum defect theory and it belongs to the category of WBEPM Theory.

Two methods have been introduced above for energy level calculations within WBEPM theory, i.e. the method by the formula of ionization energy and the three-step method with the concept of quantum defect introduced. Both methods work well on the calculation of atomic energy levels, but the orders of magnitude of the deviation are different. The absolute deviation of the first method is in the order of eV, or generally dozens of wave number (cm^{-1}) to hundreds cm^{-1}, and few may be more than hundreds cm^{-1}. The magnitude of the deviation is equivalent to that of other present methods such as MBPT, RCI, MCDF, etc., but is larger than the deviation of the second method. The absolute deviation of the second method is in the order of wave number (cm^{-1}) and most of the deviations are found to be less than 1 cm^{-1} through thousands of energy level calculations. Overall, the magnitude of its deviation is equivalent to Multichannel Quantum Defect Theory (MQDT), but we can deal with some systems that can't be or are hard to be treated by the latter, and our simple calculation can lead to good results.

Why does the first method have larger deviation than the second method? There are two reasons for that: ① the first method lacks sufficient estimation of relativistic effects. In a given iso-spectrum-level series, the contribution of the relativistic effects to energy changes with k th power of the nuclear charge, Z^k ($k \geq 4$), while in the calculation $k = 4$. δ_n of the second method is related to m^{-6}. If $k > 4$, the results of the first method may be improved. ② there is difference in the evaluations of the effects of level perturbation between two methods.

All the calculations that have been done show that the formula for energy level calculation of WBEPM Theory, i.e. Eq. (4.2.6), and the two methods for determining parameters are widely applicable to the energy level calculation with properties of universality, simplicity and accuracy for atomic systems with univalent and multivalent electrons, singly and doubly excited systems, ionic systems of high Z atom, lower and higher energy levels, unperturbed and perturbed energy levels, and all kinds of coupling types.

4.2.4 Examples

Here list only a few representative examples that we have studied.

4.2.4.1 Calculation of the Energy Levels of Carbon Group [40]

For an N-electron atom, if some particle–particle interactions are neglected, its approximate Hamiltonian can be written as [72, 73]

$$\hat{H} = \sum_i \left[\left(-\frac{1}{2}\nabla_i^2 - \frac{Z}{r_i}\right) + \sum_{i<j} \xi(r_i) l_i s_i \right]$$

$$= \sum_i \left[-\frac{1}{2}\nabla_i^2 + V(r_i) \right] + \sum_i \left[\sum_{i<j} \frac{1}{r_{ij}} - V(r_i) - \frac{Z}{r_i} \right]$$

$$+ \sum_i \xi(r_i) l_i s_i \quad (4.2.21)$$

In the above equation, the second summation term represents the part of non-central force in the repulsive interaction between electrons; the third summation term represents spin–orbit coupled magnetic interactions. In the study of atomic energy level structure, if the non-central force part of the electrostatic interactions is larger than the spin–orbit coupled magnetic interaction, LS coupling type is applicable. Lighter atoms generally belong to this type. On the contrary, if the spin–orbit coupled magnetic interaction is larger than the non-central force part of the electrostatic interactions, jj coupling type is applicable. Heavier atoms or atoms at higher energy level normally belong to this type. In addition, there is $J'l$ coupling type. $J'l$ coupling type can be seen in atoms of noble gas [73, 86].

There are three reasons that we choose the calculation of atomic energy levels of carbon group as examples: First, carbon group includes both lighter atoms and heavier atoms, and excited carbon group atoms include both LS coupling type and jj coupling type; Second, the ground-state electron configurations of carbon group include two equivalent (or identical) p electrons, so they can be regarded as the representative of multivalent-electron systems; Third, energy level series are perturbed, thus, carbon group can be regarded as one of the representatives of perturbed energy level calculations.

Tables 4.21, 4.22, 4.23, and 4.24 list, respectively, the comparison of calculated energy levels with experimental data for a spectrum-level-like series $[He]2s^2 2pnd(n \geq 3)\,^3D_1^0$ of C I, the comparison of calculated energy levels with experimental data for a spectrum-level-like series $[Ar]4s^2 3d^{10} 4pnd(n \geq 6)\,(3/2, 5/2)_3^0$ of Ge I, the comparison of calculated energy levels with experimental data for a spectrum-level-like series $[Xe]6s^2 4f^{14} 5d^{10} 6pns(n \geq 7)\,^3P_1^0$ of Pb I, and the calculated results when considering perturbation and calculated results by directly using Martin's formulas (without considering perturbation) and their comparison with experimental data for a spectrum-level-like series $[Ne]3s^2 3pnd(n \geq 3)\,^3P_0^0$ of Si I.

From this example we can see that: ① through core-valence separation the quantum defect theory considers the motion of univalent electron of an alkali metal atom in the spherical potential field of the atomic core. But for a multivalent-electron system,

4.2 Energy Level

Table 4.21 The comparison of calculated energy levels with experimental data for a spectrum-level-like series [He]$2s^2 2pnd(n \geq 3)$ $^3D_1^0$ of C I (cm^{-1}) (series limit is 90 878.3 cm^{-1})[①]

n	T_{calc}	T_{exp} [②]	$T_{calc} - T_{exp}$
3	78 318.248 4	78 318.25	−0.001 6
4	83 848.830 3	83 848.83	0.000 3
5	86 397.818 8	86 397.80	0.018 8
6	87 777.190 0	87 777.17	0.020 0
7	88 606.647 7	88 606.8	−0.152 3
8	89 143.574 4	89 143.4	0.174 4
9	89 510.840 1	89 510.9	−0.059 9
10	89 773.029 8	89 773.2	−0.170 2
11	89 966.701 2	89 966.8	−0.098 8
12	90 113.798 7		
13	90 228.138 9		
14	90 318.771 4		
15	90 391.823 9		
16	90 451.565 5		
17	90 501.043 8		
18	90 542.481 7		
19	90 577.531 5		
20	90 607.441 8		
21	90 633.170 3		
22	90 655.461 8		
23	90 674.902 3		
24	90 691.958 0		
25	90 707.003 6		
26	90 720.342 9		
27	90 732.224 6		
28	90 742.853 5		
29	90 752.399 8		
30	90 761.005 7		
31	90 768.790 7		
32	90 775.855 9		
33	90 782.287 6		
34	90 788.159 2		
35	90 793.533 9		
36	90 798.466 2		
37	90 803.003 5		
38	90 807.186 7		
39	90 811.051 8		

(continued)

Table 4.21 (continued)

n	T_{calc}	T_{exp} [2]	$T_{calc} - T_{exp}$
40	90 814.630 2		
41	90 817.949 6		
42	90 821.034 5		
43	90 823.906 3		
44	90 826.584 3		
45	90 829.085 5		
46	90 831.425 2		
47	90 833.617 0		
48	90 835.673 0		
49	90 837.604 3		
50	90 839.420 7		

Notes [1] Zheng NW, Ma DX, Yang RY, et al. J Chem Phys, 2000, 113:1681
[2] T_{exp} are taken from Fuhr JR, Martin WC, Musgrove A, et al. NIST Atomic Spectroscopic Database, Version 2.0, 1996; http://physics.gov (Select "Physical Reference Data")

Table 4.22 The comparison of calculated energy levels with experimental data for a spectrum-level-like series $[Ar]4s^2 3d^{10} 4pnd(n \geq 6)$ $(3/2, 5/2)^0_3$ of Ge I (cm^{-1}) (series limit is 65 480.60 cm^{-1})[1]

n	T_{calc}	T_{exp} [2]	$T_{calc} - T_{exp}$
6	61 268.390 0	61 268.39	−0.000 0
7	62 522.980 0	62 522.98	−0.000 0
8	63 271.309 7	63 271.31	−0.000 3
9	63 789.997 0	63 790	−0.003 0
10	64 144.081 5	64 144	0.081 5
11	64 395.707 4	64 396	−0.292 6
12	64 581.961 5	64 581.6	0.361 5
13	64 724.308 0	64 724.4	−0.092 0
14	64 835.584 7	64 835.7	−0.115 3
15	64 924.160 6	64 924.1	0.060 6
16	64 995.763 0	64 995.6	0.163 0
17	65 054.432 5	65 054.1	0.332 5
18	65 103.086 2	65 102.7	0.386 2
19	65 143.869 3	65 143.2	0.669 3
20	65 178.386 3	65 177.72	0.666 3
21	65 207.854 8	65 207.2	0.654 8
22	65 233.212 0	65 232.5	0.712 0
23	65 255.187 8	65 254.5	0.687 8

(continued)

4.2 Energy Level

Table 4.22 (continued)

n	T_{calc}	T_{exp} [2]	$T_{calc} - T_{exp}$
24	65 274.357 5	65 273.93	0.427 5
25	65 291.179 4	65 290.5	0.679 4
26	65 306.021 7	65 305.43	0.591 7
27	65 319.183 6	65 318.6	0.583 6
28	65 330.909 6	65 330.3	0.609 6
29	65 341.401 3	65 341.1	0.301 3
30	65 350.826 3	65 350.5	0.326 3
31	65 359.324 4	65 358.8	0.524 4
32	65 367.013 5	65 366.7	0.313 5
33	65 373.993 2	65 373.8	0.193 2
34	65 380.348 2	65 380.1	0.248 2
35	65 386.150 9	65 386.1	0.050 9
36	65 391.463 6	65 391.2	0.263 6
37	65 396.340 0	65 396	0.340 0
38	65 400.826 6	65 400.6	0.226 6
39	65 404.964 0	65 404.8	0.184 0
40	65 408.787 4	65 408.6	0.187 4
41	65 412.327 8	65 412.2	0.127 8
42	65 415.612 6	65 415.5	0.112 6
43	65 418.665 8	65 418.5	0.165 8
44	65 421.508 7	65 421.4	0.108 7
45	65 424.160 1	65 424.2	−0.039 9
46	65 426.636 9	65 426.5	0.136 9
47	65 428.954 1	65 428.9	0.054 1
48	65 431.125 1	65 431.1	0.025 1
49	65 433.162 0	65 433.1	0.062 0
50	65 435.075 6	65 435.1	−0.024 4
51	65 436.875 7	65 436.9	−0.024 3
52	65 438.571 0	65 438.6	−0.029 0
53	65 440.169 6	65 440.24	−0.070 4
54	65 441.678 6	65 441.79	−0.111 4
55	65 443.104 7		
56	65 444.453 8		
57	65 445.731 4	65 445.85	−0.118 6
58	65 446.942 4	65 446.99	−0.047 6
59	65 448.091 3	65 448.19	−0.098 7

(continued)

Table 4.22 (continued)

n	T_{calc}	T_{exp}[2]	$T_{calc} - T_{exp}$
60	65 449.182 4	65 449.16	0.022 4
61	65 450.219 5	65 450.29	−0.070 5
62	65 451.206 0	65 451.28	−0.074 0
63	65 452.145 3	65 452.28	−0.134 7

Notes [1] Zheng NW, Ma DX, Yang RY, et al. J Chem Phys, 2000, 113:1681
[2] T_{exp} are taken from Sugar J, Musgrove A. J Phys Chem Ref Data, 1993, 22:1213

Table 4.23 The comparison of calculated energy levels with experimental data for a spectrum-level-like series $[Xe]6s^2 4f^{14} 5d^{10} 6pns (n \geq 7)$ $^3P_1^0$ of Pb I (cm^{-1}) (series limit is 59 821.0 cm^{-1})[1]

n	T_{calc}	T_{exp}[2]	$T_{calc} - T_{exp}$
7	35 287.227 4	35 287.24	−0.012 6
8	48 686.885 7	48 686.87	0.015 7
9	53 511.611 3	53 511.34	0.271 3
10	55 719.971 6	55 720.52	−0.548 4
11	56 942.254 0	56 942.26	−0.006 0
12	57 689.254 2	57 688.73	0.524 2
13	58 179.050 6	58 179.3	−0.249 4
14	58 517.525 4	58 517.67	−0.144 6
15	58 761.169 8	58 761	0.169 8
16	58 942.370 2	58 941.8	0.570 2
17	59 080.778 4	59 080.4	0.378 4
18	59 188.881 3		
19	59 274.921 8		
20	59 344.519 7		
21	59 401.612 5		
22	59 449.026 3		
23	59 488.831 3		
24	59 522.572 8		
25	59 551.422 5		
26	59 576.282 2		
27	59 597.855 3		
28	59 616.696 5		
29	59 633.248 4		
30	59 647.867 4		
31	59 660.843 1		
32	59 672.413 0		

(continued)

4.2 Energy Level

Table 4.23 (continued)

n	T_{calc}	T_{exp}[2]	$T_{calc} - T_{exp}$
33	59 682.772 9		
34	59 692.085 9		
35	59 700.488 5		
36	59 708.095 6		
37	59 715.004 5		
38	59 721.298 1		
39	59 727.047 3		
40	59 732.313 3		
41	59 737.148 6		
42	59 741.598 9		
43	59 745.704 2		
44	59 749.499 1		
45	59 753.014 2		
46	59 756.276 3		
47	59 759.309 1		
48	59 762.133 7		
49	59 764.768 6		
50	59 767.230 5		
51	59 769.534 2		
52	59 771.692 9		
53	59 773.718 6		
54	59 775.622 0		
55	59 777.412 7		
56	59 779.099 5		
57	59 780.690 2		
58	59 782.192 1		
59	59 783.611 5		
60	59 784.954 5		

Notes [1] Zheng NW, Ma DX, Yang RY, et al. J Chem Phys, 2000, 113:1681
[2] T_{exp} are taken from Moore CE. Atomic energy levels. U.S. GPO, Washington D.C., 1971: 208

it can't be treated identically like alkali metal atoms by the idea of core-valence separation. Since equivalent (or identical) electrons can't be differentiated, Martin's formulae can only be used for univalent-electron alkali metal atoms but not for multivalent-electron carbon group, however, within WBEPM Theory every electron in a many-electron system (including univalent-electron and multivalent-electron systems) can be treated as the weakest bound electron by renaming it. Furthermore, through separation of the weakest bound electron and non-weakest bound electrons,

Table 4.24 For a spectrum-level-like series [Ne]$3s^2 3pnd (n \geq 3)\ ^3P_0^0$ of Si I, the calculated results T_{cal}^a when considering perturbation and the calculated results T_{cal}^b by directly using Martin's formulas (without considering perturbation) and their comparison with experimental data T_{exp} (cm^{-1}) (series limit is 66 035.00 cm^{-1})[①]

n	T_{exp}[②]	T_{calc}^a	$T_{calc}^a - T_{exp}$	T_{calc}^b	$T_{calc}^b - T_{exp}$
3	50 602.435	50 602.435 0	−0.000 0	50 602.435	0.000
4	56 733.369 9	56 733.369 9	−0.000 0	56 733.369 9	0.000
5	61 960.270	61 960.269 1	−0.000 9	61 960.27	0.000
6	63 123.35	63 123.360 9	0.010 9	63 123.35	0.000
7	63 863.78	63 863.756 5	−0.023 5	63 711.815 8	−151.964
8	64 358.44	64 358.429 0	−0.011 0	64 131.293 9	−227.146
9	64 703.23	64 703.293 1	0.063 1	64 460.650 2	−242.580
10	64 952.57	64 952.538 3	−0.031 7	64 723.908 6	−228.661
11	65 138.24	65 138.217 5	−0.022 5	64 934.713 9	−203.526
12	65 280.10	65 280.115 7	0.015 7	65 103.811 5	−176.289
13	65 390.91	65 390.927 6	0.017 6	65 240.040 9	−150.869
14	65 479.16	65 479.080 7	−0.079 3	65 350.502 8	−128.657
15	65 550.29	65 550.339 4	0.049 4	65 440.767 2	−109.523
16	65 608.6	65 608.750 4	0.150 4	65 515.143 7	−93.456
17	65 657.2	65 657.220 3	0.020 3	65 576.948 5	−80.251
18	65 697.6	65 697.879 9	0.279 9	65 628.733 4	−68.867
19	65 732.3	65 732.319 0	0.019 0	65 672.469 1	−59.831

Notes [①] Both T_{cal}^a and T_{cal}^b are taken from Zheng NW, Ma DX, Yang RY, et al. J Chem Phys, 2000, 113:1681
[②] T_{exp} are taken from Fuhr JR, Martin WC, Musgrove A, et al. NIST Atomic Spectroscopic Database, Version 2.0, 1996; http://physics.gov (Select "Physical Reference Data")

the motion of the weakest bound electron has been considered in the potential field of the atomic core consisting of the weakest bound electron and nucleus. Therefore, within WBEPM Theory it is reasonable to extend Martin's formulae to multivalent-electron systems. Thus, the energy levels of univalent and multivalent electrons can be treated uniformly. ② Martin's formulae are applicable to calculation of unperturbed energy levels, while it is necessary to consider perturbation for relatively common perturbed level series. Thus, it is necessary to add the second summation term in Eq. (4.2.18). For perturbed series, the differences between calculated results when considering perturbation and calculated results by directly using Martin's formulae (i.e. without considering perturbation) are shown in Table 4.23.

4.2 Energy Level

4.2.4.2 Calculation of the Energy Levels of Atom Neon and Atom Krypton [39, 42]

Both neon and krypton are noble gas, they are chosen as examples because their coupling type is not common. The energy level structure of the excited state electron configuration np^5ml of noble gas has relatively large deviation from LS coupling, so it is reasonable to use $J'l$ coupling. For this type of coupling, the angular momentums (including orbital and spin) of all electrons are first combined to be angular momentum K except for the spin s of the excited electron (i.e. the weakest bound electron), then K is coupled with s to be total angular momentum J, and the energy level can be denoted by $[K]_j^P$ [52, 72].

Based on this, then energy levels can be classified and calculated according to the spectrum-level-like series.

Part of our calculated results is listed in Tables 4.25, 4.26, 4.27, 4.28, and 4.29.

Table 4.25 The comparison of calculated energy levels with experimental data for a series $[\text{He}]2s^2 2p^5 \left({}^2P_{3/2}^0 \right) nd(n \geq 3) \, [7/2]_4^0$ of Ne I (cm^{-1}) (series limit is 173 929.6 cm^{-1})[1]

n	T_{calc}	T_{exp}[2]	$T_{\text{calc}} - T_{\text{exp}}$
1	161 590.342 0	161 590.346 2	−0.004 2
2	167 000.059 7	167 000.034	0.025 7
3	169 501.593 7	169 501.636	−0.042 3
4	170 858.478 5	170 858.465	0.013 5
5	171 675.483 6	171 675.467	0.016 6
6	172 205.120 5	172 205.130	−0.009 5
7	172 567.883 9	172 567.86	0.023 9
8	172 827.159 1	172 827.14	0.019 1
9	173 018.866 4	173 018.89	−0.023 6
10	173 164.594 7	173 164.48	0.114 7
11	173 277.952 2	173 278.08	−0.127 8
14	173 367.861 7		
15	173 440.370 7		
16	173 499.695 8		
17	173 548.849 9		
18	173 590.031 5		
19	173 624.876 2		
20	173 654.620 2		
21	173 680.212 8		
22	173 702.391 9		

(continued)

Table 4.25 (continued)

n	T_{calc}	T_{exp}②	$T_{\text{calc}} - T_{\text{exp}}$
23	173 721.738 6		
24	173 738.715 6		
25	173 753.694 5		
26	173 766.977 1		
27	173 778.810 1		
28	173 789.397 1		
29	173 798.907 1		
30	173 807.481 2		
31	173 815.238 5		
32	173 822.279 4		
33	173 828.689 6		
34	173 834.542 2		
35	173 839.899 9		
36	173 844.817 2		
37	173 849.340 9		
38	173 853.512 0		
39	173 857.366 2		
40	173 860.934 8		
41	173 864.245 3		
42	173 867.322 1		
43	173 870.186 6		
44	173 872.857 9		
45	173 875.353 0		
46	173 877.687 1		
47	173 879.873 7		
48	173 881.925 1		
49	173 883.852 0		
50	173 885.664 5		
55	173 893.292 7		
60	173 899.094 0		
65	173 903.608 2		
70	173 907.189 9		
75	173 910.079 2		
80	173 912.443 7		
85	173 914.403 2		

(continued)

4.2 Energy Level

Table 4.25 (continued)

n	T_{calc}	T_{exp} [2]	$T_{calc} - T_{exp}$
90	173 916.045 3		
95	173 917.434 9		
100	173 918.621 2		

Notes [1] Ma DX, Zheng NW, Lin X. Spectrochim Acta: Part B, 2003, 58:1625
[2] T_{exp} are taken from Fuhr JR, Martin WC, Musgrove A, et al. NIST Atomic Spectroscopic Database, Version 2.0, 1996; http://physics.gov (Select "Physical Reference Data")

Table 4.26 The comparison of calculated energy levels with experimental data for a series $[He]2s^2 2p^5 \left({}^2P^0_{3/2} \right) np (n \geq 3) [3/2]_2$ of Ne I (cm^{-1}) (series limit is 173 929.6 cm^{-1})[1]

n	T_{calc}	T_{exp} [2]	$T_{calc} - T_{exp}$
3	150 315.861 2	150 315.861 2	0.000 0
4	163 038.351 4	163 038.351 1	0.000 3
5	167 648.645 6	167 648.64	0.005 6
6	169 843.918 7	169 843.82	0.098 7
7	171 059.701 6	171 060.22	−0.518 4
8	171 803.637 1	171 803.1	0.537 1
9	172 291.824 9	172 291.4	0.424 9
10	172 629.299 6	172 630.2	−0.900 4
11	172 872.254 5	172 871.9	0.354 5
12	173 052.952 6		
13	173 190.983 3		
14	173 298.796 5		
15	173 384.611 4		
16	173 454.031 0		
17	173 510.981 6		
18	173 558.280 6		
19	173 597.992 2		
20	173 631.656 8		
21	173 660.442 8		
22	173 685.249 5		
23	173 706.777 9		
24	173 725.581 3		
25	173 742.101 0		
26	173 756.692 5		
27	173 769.644 4		
28	173 781.193 7		
29	173 791.535 8		

(continued)

Table 4.26 (continued)

n	T_{calc}	T_{exp} [2]	$T_{calc} - T_{exp}$
30	173 800.833 2		
31	173 809.222 1		
32	173 816.817 1		
33	173 823.715 4		
34	173 829.999 5		
35	173 835.740 3		
36	173 840.998 8		
37	173 845.827 3		
38	173 850.271 7		
39	173 854.371 5		
40	173 858.161 5		
41	173 861.672 1		
42	173 864.930 2		
43	173 867.959 3		
44	173 870.780 5		
45	173 873.412 4		
46	173 875.871 5		
47	173 878.172 6		
48	173 880.329 0		
49	173 882.352 5		
50	173 884.253 9		
55	173 892.235 5		
60	173 898.281 2		
65	173 902.970 0		
70	173 906.679 6		
75	173 909.664 8		
80	173 912.102 6		
85	173 914.119 2		
90	173 915.806 2		
95	173 917.231 7		
100	173 917.490 2		

Notes [1] Ma DX, Zheng NW, Lin X. Spectrochim Acta: Part B, 2003, 58:1625
[2] T_{exp} are taken from Fuhr JR, Martin WC, Musgrove A, et al. NIST Atomic Spectroscopic Database, Version 2.0, 1996; http://physics.gov (Select "Physical Reference Data")

4.2 Energy Level

Table 4.27 The comparison of calculated energy levels with experimental data for a series $[\text{He}]2s^22p^5\left(^2P^0_{3/2}\right)ns(n \geq 3)\,[3/2]^0_1$ of Ne I (cm^{-1}) (series limit is 173 929.6 cm^{-1})[①]

n	T_{calc}	T_{exp}[②]	$T_{\text{calc}} - T_{\text{exp}}$
3	134 459.224 3	134 459.287 1	−0.062 8
4	158 796.568 1	158 795.992 4	0.575 7
5	165 911.658 4	165 912.782	−1.123 6
6	168 967.757 5	168 967.353	0.404 5
7	170 557.509 4	170 557.050	0.459 4
8	171 489.215 1	171 489.485	−0.269 9
9	172 081.863 4	172 080.916	0.947 4
10	172 482.107 3	172 481.86	0.247 3
11	172 765.069 6	172 764.57	0.499 6
12	172 972.480 0	172 972.36	0.120 0
13	173 129.028 9	173 128.78	0.248 9
14	173 250.085 0		
15	173 345.621 0		
16	173 422.337 7		
17	173 484.872 2		
18	173 536.516 8		
19	173 579.660 8		
20	173 616.072 4		
21	173 647.083 0		
22	173 673.710 2		
23	173 696.742 9		
24	173 716.800 2		
25	173 734.373 2		
26	173 749.856 1		
27	173 763.567 5		
28	173 775.767 8		
29	173 786.671 2		
30	173 796.455 1		
31	173 805.267 7		
32	173 813.233 4		
33	173 820.457 5		
34	173 827.029 1		
35	173 833.024 6		
36	173 838.509 4		
37	173 843.539 9		
38	173 848.164 8		

(continued)

Table 4.27 (continued)

n	T_{calc}	T_{exp} [2]	$T_{calc} - T_{exp}$
39	173 852.426 7		
40	173 856.362 6		
41	173 860.004 9		
42	173 863.382 1		
43	173 866.519 3		
44	173 869.438 7		
45	173 872.160 1		
46	173 874.700 9		
47	173 877.076 7		
48	173 879.301 6		
49	173 881.388 1		
50	173 883.347 4		
55	173 891.558 3		
60	173 897.762 1		
65	173 902.563 4		
70	173 906.355 2		
75	173 909.401 8		
80	173 911.886 5		
85	173 913.939 4		
90	173 915.655 0		
95	173 917.103 4		
100	173 918.337 4		

Notes [1] Ma DX, Zheng NW, Lin X. Spectrochim Acta: Part B, 2003, 58:1625
[2] T_{exp} are taken from Fuhr JR, Martin WC, Musgrove A, et al. NIST Atomic Spectroscopic Database, Version 2.0, 1996; http://physics.gov (Select "Physical Reference Data")

Table 4.28 The comparison of calculated energy levels with experimental data for a series $[Ar]4s^2 3d^{10} 4p^5 \left({}^2P^0_{3/2} \right) nd (n \geq 4) \, {}^2[3/2]^0_1$ of Kr I (cm^{-1}) (series limit is 112 912.40 cm^{-1})[1]

n	T_{exp} [2]	T_{calc}	$T_{calc} - T_{exp}$
4	99 646.208 6	99 646.208 6	−0.000 0
5	105 648.428 7	105 648.428 7	0.000 0
6	108 258.750 7	108 258.750 4	−0.000 3
7	109 688.751 1	109 688.752 6	0.001 5
8	110 514.090 1	110 514.089 5	−0.000 6
9	111 154.3	111 154.302 1	0.0
10	111 520.2	111 520.172 3	−0.0
11	111 786.1	111 786.148 0	0.0

(continued)

4.2 Energy Level

Table 4.28 (continued)

n	T_{exp}[2]	T_{calc}	$T_{\text{calc}} - T_{\text{exp}}$
12	111 983.3	111 983.286 6	−0.0
13	112 133.2	112 133.179 0	−0.0
14	112 249.7	112 249.711 5	0.0
15	112 342.0	112 342.055 9	0.1
16	112 416.4	112 416.450 2	0.1
17	112 477.2	112 477.250 9	0.1
18	112 527.5	112 527.572 6	0.1
19	112 569.7	112 569.688 2	−0.0
20	112 605.2	112 605.287 6	0.1
21	112 635.6	112 635.647 4	0.0
22	112 661.8	112 661.746 7	−0.1
23	112 684.3	112 684.346 6	0.0
24	112 703.95	112 704.045 3	0.20
25	112 721.23	112 721.318 9	0.09
26	112 736.46	112 736.549 4	0.09
27	112 749.97	112 750.046 8	0.08
28	112 762.00	112 762.064 3	0.06
29	112 772.75	112 772.810 5	0.06
30	112 782.40	112 782.458 5	0.06
31	112 791.10	112 791.153 1	0.05
32	112 798.97	112 799.015 9	0.05
33	112 806.12	112 806.149 6	0.03
34	112 812.59	112 812.641 6	0.05
35	112 818.53	112 818.566 7	0.04
36	112 823.96	112 823.989 1	0.03
37	112 828.93	112 828.963 9	0.03
38	112 833.50	112 833.539 1	0.04
39	112 837.74	112 837.756 5	0.02
40	112 841.62	112 841.652 3	0.03
50	112 868.6	112 868.393 0	−0.21
60	112 883.4	112 882.702 3	−0.70

Notes [1] Zheng NW, Zhou T, Yang RY, et al. Chem Phys, 2000, 258:37
[2] T_{exp} are taken from Sugar J, Musgrove A. J Phys Chem Ref Data, 1991, 20:859

Table 4.29 The comparison of calculated energy levels with experimental data for a series $[Ar]4s^2 3d^{10} 4p^5 \left({}^2P^0_{3/2} \right) ns(n \geq 5)\, {}^2[3/2]^0_2$ of Kr I (cm^{-1}) (series limit is 112 912.40 cm^{-1})[①]

n	T_{exp}[②]	T_{calc}	$T_{calc} - T_{exp}$
5	79 971.732 1	79 971.731 9	−0.000 2
6	99 626.875 3	99 626.875 6	0.000 3
7	105 647.448 2	105 647.447 6	−0.000 6
8	108 324.977 9	108 324.984 4	0.006 5
9	109 751.959 3	109 751.956 9	−0.002 4
10	110 608.353 7	110 608.360 2	0.006 5
11	111 154.390 0	111 154.356 8	−0.033 2
12	111 527.820 0	111 527.841 9	0.021 9
13		111 793.964 7	
14		111 990.222 0	
15		112 139.086 8	
16		112 254.675 8	
17		112 346.215 8	
18		112 419.943 9	
19		112 480.198 8	
20		112 530.074 3	
21		112 571.824 4	
22		112 607.123 0	
23		112 637.234 0	
24		112 663.126 3	
25		112 685.552 7	
26	112 705.100 0	112 705.105 4	0.005 4
27	112 722.260 0	112 722.255 1	−0.004 9
28	112 737.370 0	112 737.380 2	0.010 2
29	112 750.740 0	112 750.787 1	0.047 1
30	112 762.740 0	112 762.726 6	−0.013 4
31	112 773.410 0	112 773.405 3	−0.004 7
32	112 783.010 0	112 782.994 7	−0.015 3
33	112 791.610 0	112 791.638 1	0.028 1
34	112 799.470 0	112 799.455 8	−0.014 2
35	112 806.560 0	112 806.549 9	−0.010 1
36	112 813.020 0	112 813.006 9	−0.013 1
37	112 818.920 0	112 818.901 0	−0.019 0
38	112 824.320 0	112 824.295 6	−0.024 4
39	112 829.250 0	112 829.245 7	−0.004 3
40	112 833.810 0	112 833.798 9	−0.011 1

(continued)

4.2 Energy Level

Table 4.29 (continued)

n	T_{exp} [2]	T_{calc}	$T_{calc} - T_{exp}$
50	112 864.520 0	112 864.507 2	−0.012 8
60	112 880.490 0	112 880.503 6	0.013 6

Notes [1] Zheng NW, Zhou T, Yang RY, et al. Chem Phys, 2000, 258:37
[2] T_{exp} are taken from Sugar J, Musgrove A. J Phys Chem Ref Data, 1991, 20:859.
Remark From NIST Atomic Spectra Database, Version 3, we found the experimental data for energy levels of $n = 13 - 20$ and series limit (cm^{-1}) for a series $[Ar]4s^2 3d^{10} 4p^5 (^2P_{3/2}) ns^2 \, [3/2]_2^0$ of Kr I. They are, respectively, $n = 13, 14, 15, 16, 17, 18, 19, 20$, $T_{exp} = 111\,793.1,\, 111\,990.0,\, 112\,139.4,\, 112\,254.7,\, 112\,346.1,\, 112\,420.0,\, 112\,480.3,\, 112\,530.1$. Series limit is 112 914.433. The original data in Version 3 came from Saloman EB. J Phys Chem Ref Data, 2007, 36:215

4.2.4.3 Calculation of Autoionizing Levels of Doubly Excited States for Atom Carbon [47]

This example has the following properties: ① double-electron excitation. The ground-state electron configuration of Ca I is $[Ar]4s^2$. The electron configuration of 30 series that we calculated for is $[Ar]3dnl$ in which two 4s electrons are excited to 3d and nl, respectively, so this is double-electron excitation. ② Even-parity state. Atoms have center of symmetry, and under inversion operation, the wave functions either keep the sign or change the sign. The former is called even-parity state and the latter is called odd-parity state. The one with "0" on the upper right corner of the symbol of energy level is the odd-parity state. How to find the parity for a state? We can multiply the orbital parity of every electron (odd × odd = even × even = even, odd × even = odd), or take the sum of l for every electron. Parity is related to both energy level perturbation and transition, so parity operator is important in atomic spectrum [73, 87]. ③ Autoionizing state. In atomic energy levels, if the energy of a discrete level a in a level series is close to that of a continuous level b in another series, there is configuration interaction between them. Thus, when the atom is excited to the discrete a state, there is certain probability that it will automatically transit to b state, making the atom ionized. This special discrete energy level is called autoionizing energy level [72].

Part of the results is listed in Tables 4.30 and 4.31.

4.2.4.4 Calculation of Energy Levels of Atom Strontium [48]

The properties of this example are: ① relatively high atomic number of Sr I; ② double-electron excitation; ③ for perturbed energy levels, two methods with and without considering perturbations have been used for calculation in order to evaluate the effect of perturbation on the magnitude of deviation.

Part of the results is listed in Tables 4.32 and 4.33.

Table 4.30 The comparison of calculated energy levels with experimental data and values of R/MQDT for a series $[Ar]3d_{3/2}nd_{5/2} J = 1$ of Ca I (cm^{-1}) (series limit is 62 956.15 cm^{-1})[①]

n	T_{exp}[②]	$T_{WBEPM, calc}$	R/MQDT value[③]
6	58 807.9	58 807.9	58 801.5
7	60 062.6	60 062.6	60 059.2
8	60 820.4	60 820.4	60 819.0
9	61 328.2	61 328.2	61 328.0
10	61 661.5	61 661.5	61 661.4
11	61 910.7	61 910.7	61 910.7
12	62 088.6	62 088.6	62 088.6
13	62 224.3	62 224.2	62 224.4
14	62 329.9	62 330.0	62 330.2
15	62 414.3	62 414.3	62 414.7
16	62 483.1	62 482.5	62 483.6
17	62 535.8	62 538.7	62 535.9
18	62 583.5	62 585.4	62 584.2
19	62 623.5	62 624.7	62 623.8
20	62 656.6	62 658.1	62 657.4
21	62 685.4	62 686.7	62 686.4
22	62 710.3	62 711.4	62 710.8
23	62 731.8	62 732.9	62 732.4
24	62 750.7	62 751.7	62 751.2
25	62 767.5	62 768.2	62 767.7
26	62 781.9	62 782.7	62 782.3
27	62 795.0	62 795.7	62 795.4
28	62 806.5	62 807.3	62 806.9
29	62 817.0	62 817.6	62 817.3
30	62 826.2	62 826.9	62 826.6
31	62 834.8	62 835.3	62 835.1
32	62 842.2	62 843.0	62 842.6
33	62 849.5	62 849.9	62 849.7
34		62 856.2	
35	62 861.7	62 861.9	62 861.7
36	62 866.7	62 867.2	62 867.0
37		62 872.1	
38		62 876.5	
39		62 880.6	
40		62 884.5	
41		62 888.0	
42		62 891.3	

(continued)

4.2 Energy Level

Table 4.30 (continued)

n	T_{exp}[2]	T WBEPM, calc	R/MQDT value[3]
43		62 894.3	
44		62 897.1	
45		62 899.8	
46		62 902.2	
47		62 904.5	
48		62 906.7	
49		62 908.7	
50		62 910.7	
51		62 912.5	
52		62 914.1	
53		62 915.7	
54		62 917.2	
55		62 918.7	
60		62 924.7	
65		62 929.4	
70		62 933.2	
75		62 936.2	
80		62 938.6	
85		62 940.6	
90		62 942.3	
100		62 945.0	

Notes [1] Ma DX, Zheng NW, Fan J. J Phys Chem Ref Data, 2004, 33:1013
[2] T_{exp} are taken from Assimopoulos S, Bolovinos A, Jimoyiannis A, et al. J Phys B: At Mol Opt Phys, 1994, 27: 2471
[3] The values of R/MQDT are taken from Assimopoulos S, Bolovinos A, Jimoyiannis A, et al. J Phys B: At Mol Opt Phys, 1994, 27:2471

Table 4.31 The comparison of calculated energy levels with experimental data and values of R/MQDT for a series [Ar]$3d_{5/2}nd_{3/2} J = 2$ of Ca I (cm^{-1}) (series limit is 63 016.84 cm^{-1})[1]

n	T_{exp}[2]	T WBEPM, calc	R/MQDT value[3]
5	57 578.9	57 578.9	57 678.2
6	59 363.5	59 363.7	59 396.7
7	60 378.0	60 376.7	60 399.8
8	61 023.7	61 025.6	61 035.8
9	61 457.0	61 456.3	61 464.0
10		61 687.5	
11	61 986.9	61 986.8	61 989.0
12		62 155.3	

(continued)

Table 4.31 (continued)

n	$T_{\exp}$[2]	$T_{\text{WBEPM, calc}}$	R/MQDT value[3]
13	62 286.6	62 286.3	62 287.2
14	62 389.8	62 390.0	62 390.0
15	62 472.1	62 473.4	62 471.8
16	62 544.5	62 541.3	62 545.4
17	62 600.5	62 597.4	62 601.3
18	62 645.5	62 644.3	62 646.2
19	62 684.3	62 683.7	62 684.2
20		62 717.3	
21	62 746.5	62 746.1	62 747.2
22	62 771.0	62 770.9	62 771.4
23	62 793.0	62 792.5	62 793.3
24	62 812.1	62 811.4	62 812.5
25	62 828.0	62 828.0	62 828.4
26		62 842.6	62 843.1
27		62 855.7	62 856.1
28		62 867.3	62 867.7
29		62 877.7	
30		62 887.1	
31		62 895.6	
32		62 903.2	
33		62 910.2	
34		62 916.5	
35		62 922.3	
36		62 927.6	
37		62 932.5	
38		62 937.0	
39		62 941.1	
40		62 944.9	
41		62 948.5	
42		62 951.8	
43		62 954.8	
44		62 957.7	
45		62 960.3	
46		62 962.8	
47		62 965.1	
48		62 967.3	
49		62 969.3	

(continued)

Table 4.31 (continued)

n	T_{exp}[2]	$T_{WBEPM, calc}$	R/MQDT value[3]
50		62 971.2	
51		62 973.0	
52		62 974.7	
53		62 976.3	
54		62 977.8	
55		62 979.3	
60		62 985.4	
65		62 990.1	
70		62 993.8	
75		62 996.8	
80		62 999.3	
90		63 003.0	
100		63 005.7	

Notes [1] Ma DX, Zheng NW, Fan J. J Phys Chem Ref Data, 2004, 33:1013
[2] T_{exp} are taken from Assimopoulos S, Bolovinos A, Jimoyiannis A, et al. J Phys B: At Mol Opt Phys, 1994, 27:2471
[3] The values of R/MQDT are taken from Assimopoulos S, Bolovinos A, Jimoyiannis A, et al. J Phys B: At Mol Opt Phys, 1994, 27:2471

Table 4.32 The comparison of calculated energy levels with experimental data for a series Sr[Kr]$4d_{3/2}nd_{3/2} J = 1$ (cm^{-1}) (series limit is 60 488.09 cm^{-1})[1]

n	T_{exp}[2]	$T_{WEBPM, calc}$	$T_{exp} - T_{WBEPM, calc}$
6	54 139.2	54 139.2	0.0
7	56 366.9	56 366.9	0.0
8	57 683.8	57 683.8	0.0
9	58 378.1	58 378.0	0.1
10	58 859.0	58 859.4	−0.4
11	59 193.6	59 193.1	0.5
12	59 433.7	59 433.9	−0.2
13	59 613.2	59 613.3	−0.1
14	59 750.6	59 750.6	0.0
15	59 859.5	59 857.9	1.6
16	59 944.9	59 943.4	1.5
17	60 013.9	60 012.7	1.2
18	60 070.6	60 069.5	1.1
19	60 117.6	60 116.7	0.9
20	60 157.4	60 156.4	1.0

(continued)

Table 4.32 (continued)

n	T_{exp} [2]	$T_{WEBPM, calc}$	$T_{exp} - T_{WBEPM, calc}$
25	60 284.4	60 284.0	0.4
30	60 350.1	60 350.0	0.1
35	60 388.6	60 388.4	0.2
40	60 412.9	60 412.8	0.1
45	60 429.3	60 429.2	0.1
50	60 440.9	60 440.8	0.1
55	60 449.3	60 449.3	0.0
60	60 455.7	60 455.7	0.0
65	60 460.6	60 460.6	0.0
70	60 464.5	60 464.5	0.0
80	60 470.2	60 470.1	0.1

Notes [1] Zhang TY, Zheng NW, Ma DX. Phys Scr, 2007, 75:763
[2] T_{exp} are taken from Goutis S, Aymar M, Kompitsas M, Camus P. J Phys B: At Mol Opt Phys, 1992, 25:3433. And the perturbed energy levels are 5p6p $^1P_1$ (57 305.9 cm^{-1}) and 5p6p $^3D_1$ (58 009.8 cm^{-1})

Table 4.33 The calculated energy levels with and without considering perturbations and their comparison with experimental data for a series Sr[Kr]$4d_{5/2}nd_{5/2} J = 1$ (cm^{-1}) (series limit is 60 768.43 cm^{-1})[1]

n	T_{exp} [2]	$T_{p, cal}$ [3]	$T_{unp, cal}$ [4]
6	54 469.5	54 469.5	54 473.6
7	56 676.0	56 675.9	56 654.7
8	57 836.4	57 836.4	57 869.4
9	58 642.0	58 641.8	58 630.1
10	59 137.4	59 137.2	59 130.0
11	59 469.8	59 470.6	59 473.5
13	59 899.8	59 900.3	59 900.2
14	60 035.3	60 034.6	60 038.0
15	60 141.6	60 141.3	60 145.2
25	60 565.1	60 565.2	60 566.8
35	60 669.2	60 669.1	60 669.7
45	60 709.9	60 709.8	60 710.0
55	60 729.8	60 729.7	60 729.9

(continued)

Table 4.33 (continued)

n	T_{exp} ②	$T_{p,\text{cal}}$ ③	$T_{\text{unp,cal}}$ ④
65	60 741.0	60 741.0	60 741.1
75	60 748.0	60 748.0	60 748.0
85	60 752.6	60 752.6	60 752.6

Notes ① Zhang TY, Zheng NW, Ma DX. Phys Scr, 2007, 75:763
② T_{exp} are taken from Goutis S, Aymar M, Kompitsas M, Camus P. J Phys B: At Mol Opt Phys, 1992, 25:3433. And the perturbed energy levels are 5p6p $^1P_1$ (57 305.9 cm^{-1}) and 5p6p $^3D_1$ (58 009.8 cm^{-1}), 5p6p $^3P_1$ (59 011.3 cm^{-1}) and 5p6p $^3S_1$ (59 796.3 cm^{-1})
③ $T_{p,cal}$ are the results calculated using WBEPM Theory with perturbations considered
④ $T_{unp,cal}$ are the results calculated using Martin's formulas without considering perturbations

4.3 Calculation of Oscillator Strength, Transition Probability and Radiative Lifetime [88–104]

4.3.1 Introduction

The theory of bound state is mainly interested in the eigenvalues and eigenstates of discrete energies of a given system and quantum transitions between them [105]. So theoretical study of quantum transition (transition probability, oscillator strength, radiative lifetime, etc.) is very important. It not only provides basic information for related structures, but also provides strong judgments for the reliability and accuracy of all kinds of approximate theoretical methods by comparing theory with experimental data. In addition, the study of transition probability, oscillator strength, and radiative lifetime is essential for the practical applications in the field of high technology such as chemistry, astrophysics, physics, laser, nuclear fusion, plasma, etc. For instance, spectroscopic analysis in chemistry is necessary for structure detections, and the discovery of new working-laser materials requires the study of quantum transition properties, etc. Therefore, the study of transition probability, oscillator strength, and radiative lifetime always receives high attention from both sides of experiment and theory. The main theoretical methods for research in this area are the Multiconfiguration Hartree–Fock (MCHF) method, Multi-Configuration Dirac–Fock method (MCDF), Configuration Interaction (CI) method, relativistic Hartree–Fock (relativistic HF) method, Many-Body Perturbation Theory (MBPT), Quantum Defect Theory (QDT), R matrix method, Model Potential (MP) method, Coulomb Approximation (CA), etc. [106–132]. Many progresses have been made on the research, but overall, problems still exist in the following three aspects: ① research work focuses on lower excited states and lower ionization states, while work on higher excited states and higher ionization states is missing due to the difficulties in the study; there are relatively more results for lower Z elements, especially elements with fewer valence electrons, while the results for higher Z elements and elements with more valence electrons are rare or absent. ② There are more studies on transitions between spectral terms and fewer studies on the transition between energy levels, but in practical applications transitions between energy levels are much more important than transitions between spectral terms. ③ Calculation is complicated.

In the past few years, we have widely and systematically studied transition probability, oscillator strength, and radiative lifetime of many-electron atomic or ionic systems using WBEPM Theory. Studies involve transition probability between energy levels of C I–C IV [88], transition probability for titanium ion (TiIII and TiIV) [89], transition probability for nitrogen (N I–N V) [90], transition probability for oxygen (O I–O VI) [91], transition probability for fluorine atom [92], transition probability between electron configurations of lower states for alkali metal atoms [93], transition probability between electron configurations of lithium atom and lithium-like ions [94], transition probability between energy levels of boron atom (B I) [95], transition probability for Cu I, Ag I and Au I [96], transition probability between spectral terms of Be I, Be II, Mg I and Mg II [97], resonance transition probability and radiative lifetime between spectral terms of nitrogen atom [98], transition probability and radiative lifetime for carbon atom and oxygen atom [99], transition probability for Ne II [92], transition probability for Al II [101], oscillator strength and transition probability for magnesium-like ions [102], spin-allowed transitions between energy levels above 3s3p state in Si III [103].

In comparison with all other theoretical methos, WBEPM Theory has four properties. First, we have studied plenty of systems many of which have different electron configurations, spectral terms and energy levels, but the calculated results for transition probability, oscillator strength, and radiative lifetime between electron configurations, spectral terms and energy levels are generally in agreement with acceptable values, or experimental results, or results from all other theoretical methods. This indicates that WBEPM Theory is effective and reliable in the study of quantum transitions. Second, lots of elements have been studied and involved state-state transitions are different (including transitions between electron configurations, spectral terms and energy levels), but the theorem, method and program of calculation remains the same, indicting the consistence of the theory, which is different from ab initio methods. In ab initio methods, the choice of base sets and the choice of configuration state functions are closely correlated with accuracy of the results and the complexity of the calculations, and it is common to choose different base sets and configuration state functions which causes inconsistency of the results. Third, the published work includes study on both lower excited states and higher excited states, and both lower ionization states and higher ionization states. Fourth, the calculation process of WBEPM Theory is very simple.

When using WBEPM Theory to study transition problems, three points must be paid attention to: ① Sometimes it is necessary to adjust the number of nodes. ② When using associated equations to determine parameters, the accuracy of radial expectation values is closely correlated with the accuracy of the results. So it is important to choose a method with $\langle r \rangle$ close to a real value. ③ Relativistic effects should be corrected for higher ionization states. Above problems will be discussed when dealing with specific objects in the following section.

4.3.2 Theory and Method for Calculation

The formulae for calculating the probability of transition between electron configurations A'_{fi}, the probability of transition between spectral terms A''_{fi}, and the probability of transition between energy levels A'''_{fi} of two states f and i are as follows [104, 133]:

$$A'_{fi} = \frac{4}{3}\alpha^3(E_f - E_i)^3 |\langle n_f l_f | r | n_i l_i \rangle|^2 \frac{l_>}{2l_i + 1} \tag{4.3.1}$$

$$A''_{fi} = \frac{4}{3}\alpha^3(E_f - E_i)^3 |\langle n_f l_f | r | n_i l_i \rangle|^2 \\ \times (2L_i + 1) l_> W^2(l_i L_i l_f L_f; L_c 1) \tag{4.3.2}$$

$$A'''_{fi} = \frac{4}{3}\alpha^3(E_f - E_i)^3 |\langle n_f l_f | r | n_i l_i \rangle|^2 \\ \times (2L_f + 1)(2L_i + 1)(2J_i + 1)l_> \\ \times W^2(l_i L_i l_f L_f; L_c 1) \times W^2(L_i J_i L_f J_f; S1) \tag{4.3.3}$$

In above three equations, α is the fine-structure constant, $l_> = \max(l_f, l_i)$, L_c is total angular momentum of the atomic core, $W(abcd; ef)$ is Racah coefficient [134]. For transition between excited states, the above three equations can be directly used. While for resonance transition, the right side of the above three equations need to be multiplied by the number of electrons on the orbital l and a scale factor [97].

The relation between oscillator strength f_{if}, radiative lifetime τ_f and transition probability is as follows:

$$f_{if} = 1.499 \times 10^{-8} \lambda^2 \frac{g_f}{g_i} A_{fi} \tag{4.3.4}$$

where g is lande factor.

$$\tau_f = \frac{1}{\sum_i A_{fi}} \tag{4.3.5}$$

τ_f is radiative lifetime of state f, $\sum_i$ represents the summation of all states i which have lower energy than state f and between which and state f transitions can occur.

From above formulae, calculation of transition probability, oscillator strength, and radiative lifetime requires the values of E_f and E_i as well as calculation of transition matrix elements. Among them, the latter is the key of the problem.

How to determine E_f and E_i?

For states f and i, either both are excited states or one is excited state and another is ground state. The negative of the energy of the weakest bound electron at states f and

i is equal to the difference between series limit and energy of the state. The relativistic correction caused by mass-velocity effect is very small and if the relativistic effect of spin-orbital coupling is also very small, the relativistic effect can be neglected. We rewrite Eq. (4.1.13) as

$$\varepsilon = -\frac{Z'^2}{2n'^2} \approx \varepsilon_\mu^0 \tag{4.3.6}$$

Then

$$E_{spec} = -\varepsilon_\mu^0 \approx \frac{Z'^2}{2n'^2} \text{ (Neglecting electron correlation and relativistic effect)} \tag{4.3.7}$$

$$E_{spec} \approx -\varepsilon_\mu^0 \approx \frac{Z'^2}{2n'^2} \text{ (Without neglecting electron correlation and relativistic effect)} \tag{4.3.8}$$

Experimentally,

$$E_{spec} = I_{exp} = T_{limit} - T(n) \tag{4.3.9}$$

At present, there are lots of very accurate experimental data for energy levels of atoms and ions, and they are collected in relevant database, handbook, and collections. Data published by National Institute and Standards and Technology (NIST database) are very complete [135]. In our previous studies, for convenience, E_{spec} for both state f and state i are taken from NIST database. If lacking data, one can use the method for energy level calculation in WBEPM Theory that was stated in Sect. 4.2 to calculate E_f and E_i. This method is very simple and very accurate, and by comparing with experimental data, the calculated values by our method have very small absolute deviations from experimental data.

About calculation of transition matrix elements, in Chap. 3 analytical formula for radial wave function and formula for matrix elements are given for the weakest bound electron. Based on this, the matrix elements can be calculated in principle. However, in practical calculation process, parameters Z', n' and l' that are included are required to be determined (since $n' = n + d$, $l' = l + d$, there are only two parameters, i.e. Z' and d that actually need to be determined.). To avoid complicated calculations that other methods especially ab initio methods encountered, we have proposed a very simple method to determine parameters, which is called associated equations.

The associated equations consist of the energy formula of the weakest bound electron, i.e. Eq. (4.3.7) (note that "≈" has been written as "=".) and formula of expectation value $\langle r \rangle$ of radial position operator, i.e. Eq. (3.3.3).

4.3 Calculation of Oscillator Strength, Transition Probability ...

$$\begin{cases} E_{spec} = \frac{Z'^2}{2n'^2} \\ \langle r \rangle = \frac{3n'^2 - l'(l'+1)}{2Z'} \end{cases} \quad (4.3.10)$$

where E_{spec} is obtained from NIST value, $\langle r \rangle$ can be calculated by many theoretical methods at present, for instance numerical Coulomb approximation (NCA), Roothann–Hartree–Fock (RHF) method, multiconfiguration Hartree–Forck (MCHF) method, Hartree–Slater (HS) method, Hartree–Kohn–Sham (HKS) method, time-dependent Hartree–Fock (TDHF) method, self-interaction-corrected local-spin-density (SIC-LSD) method, etc. [88, 94, 103].

For simplicity and convenience of calculation, in the previous work we generally obtained the value of $\langle r \rangle$ by NCA method [136, 137]. NCA method is simple and easy to use and gives very good results for excited states. But for ground state, the deviation between $\langle r \rangle$ obtained by NCA and true $\langle r \rangle$ is relatively large [138, 139], causing relatively large errors in calculation of transitions, thus, in the calculation of resonance transition probability NCA method is replaced by RHF method to obtain $\langle r \rangle$ of ground state [97, 140, 141].

Substituting E_{spec} and $\langle r \rangle$ of states f and i into associated Eq. (4.3.10), we can obtain parameters n'_f, l'_f and Z'_f as well as n'_i, l'_i and Z'_i, and substituting these parameters into the formula for calculation of matrix elements, we can get the values of matrix elements $\langle n_f l_f | r | n_i l_i \rangle$ in Eqs. (4.3.1)–(4.3.3). Thus, transition probability, oscillator strength, and radiative lifetime can be computed one by one.

From the above description, it is not hard to see that the use of associated equations to determine parameters makes the calculation of transition probability, oscillator strength, and radiative lifetime easy and convenient, and generally makes results good as well.

The deviation mainly comes from two aspects: ① E_{spec} is taken from experimental value, naturally including relativistic effects, while the formula of $\langle r \rangle$ comes from non-relativistic theoretical methods, so two can't match perfectly when the relativistic effect is relatively large, which results in deviation, but this deviation generally is not primary. ② The choice of $\langle r \rangle$ is closely correlated with the accuracy for the calculation of transitions. The closer to true $\langle r \rangle$ the selected $\langle r \rangle$ is, the better the results are.

Tables 4.34 and 4.35 list the values of $\langle r \rangle$ determined by HKS method and model potential method of Theodosiou as well as the values of 3s-3p and 3p-3d transition probabilities for sodium [93, 138].

From Tables 4.34 and 4.35, we can see that the values of $\langle r \rangle$ obtained by two methods are not completely identical, and the difference is bigger for lower energy state (3s); using $\langle r \rangle$ obtained by different theoretical methods to calculate transition probability, the results are different as well. Data in Table 4.35 show that the results of WBEPM Theory are better than that of HKS method. It indicates how important the choice of $\langle r \rangle$ is to the calculation of quantum transitions. The closer to true $\langle r \rangle$ the selected $\langle r \rangle$ is, the better the results of calculation of transitions are.

Celik et al. also used two different methods of NCA and NRHF (numerical nonrelativistic Hartree-Forck) within WBEPM Theory to determine parameters

Table 4.34 $\langle r \rangle$ of 3s, 3p and 3d orbital obtained by HKS method and model potential method

Orbital	$\langle r \rangle$ value	
	HKS method	Model potential method
3s	4.277	3.971
3p	5.972	5.720
3d	10.425	10.408

Notes [1] Theodosiou CE. Phys Rev A, 1984, 30:2881
[2] Zheng NW, Wang T, et al. J Phys Soc JPN, 1999, 68:3859

Table 4.35 Comparison of some transition probabilities ($\times 10^8$ Hz) of sodium obtained by determining $\langle r \rangle$ using two different methods[1]

Transition	HKS method[2]		WBEPM theory		Acceptable value
	(1)	(2)	(1)	(2)	
3s – 3p	0.677 466	0.710 326	0.592 464	0.620 348	0.629
3p–3d	0.530 198	0.551 147	0.474 019	0.496 063	0.495

Data source Zheng NW, Wang T, et al. J Phys Soc JPN, 1999, 68:3859
Notes [1] In the table (1) represents calculated values without correction of nodes number, (2) represents calculated values with correction of nodes number
[2] Theodosiou CE. Phys Rev A, 1984, 30:2881

through associated equations and made comparisons for some calculated transition probabilities of p-d and d-p transitions of excited nitrogen atom [142].

In our previous studies, the calculated values for transition (including transition probability, oscillator strength, and radiative lifetime) between lower excited states have relatively large deviations comparing with acceptable values or measurements or results by other theoretical methods, however, the calculated values for transition between higher excited states are clearly better than that of other theoretical methods, and are in a good agreement with acceptable values or measurements. The key reason is still directly related with the radial expectation value $\langle r \rangle$.

One more thing that needs to mention is that from Eq. (4.3.5) we know that the calculation of radiative lifetime involves the summation of all states i which have lower energy than state f and between which and state f transitions can occur. Only when all the transition probabilities in the summation are correct can τ_f be correct. It is not very easy to accomplish this because although the results of some transition probabilities are relatively correct, the results of other transition probabilities may be not correct enough which would affect the accuracy of τ_f. This is one of the difficulties that all theoretical methods face at present. So the difference between literature data from different theoretical methods can be relatively large.

4.3.3 Examples

Using WBEPM Theory, we have systematically studied the transition properties of many representative atoms and ions, and we not only provided results of comparison with known experimental data or known theoretical values, but also predicted plenty of transition data between high excited states and transition data for ions at relatively higher ionization states. These results indicate that WBEPM Theory has four properties in the study of transitions, i.e. accuracy, consistency, university and simplicity. Part of the results is listed in Tables 4.36, 4.37, 4.38, 4.39, 4.40, 4.41, and 4.42. Readers can further investigate relevant literature if interested.

Our work has interested experimental and theoretical physicists [55, 143–158]. Celik et al. have used WBEPM Theory for the following studies: ① theoretical calculation of transition probability of some excited p-d transitions for nitrogen atom; ② calculation of transition probability for oxygen atoms; ③ comparison of transition probabilities of some p-d and d-p transitions for an excited nitrogen atom which are calculated by different parameters within WBEPM Theory; ④ calculation of transition probability between spectral lines for lithium atom. They also pointed

Table 4.36 Transition probability between energy levels of magnesium-like ions (Al II, Si III, P IV, S V, Cl VI, Ar VII)

Ions	Transition	Transition probability (10^8 s^{-1})		
		WBEPM theory	NIST*	Others
Al II[①]	$4s\,^3S_1 - 4p\,^3P^0_0$	0.586 701	0.58	0.570 47[a]
	$4s\,^3S_1 - 4p\,^3P^0_1$	0.588 285	0.58	0.572 38[a]
	$4s\,^3S_1 - 4p\,^3P^0_2$	0.591 624	0.59	0.576 24[a]
	$5s\,^3S_1 - 4p\,^3P^0_0$	0.104 924	0.11	
	$5s\,^3S_1 - 4p\,^3P^0_1$	0.314 672	0.34	
	$5s\,^3S_1 - 4p\,^3P^0_2$	0.524 094	0.57	
	$6s\,^3S_1 - 4p\,^3P^0_0$	0.050 694	0.043	
	$6s\,^3S_1 - 4p\,^3P^0_1$	0.151 999	0.13	
	$6s\,^3S_1 - 4p\,^3P^0_2$	0.253 032	0.21	
	$8s\,^3S_1 - 5p\,^3P^0_0$	0.009 988	0.008 5	
	$8s\,^3S_1 - 5p\,^3P^0_1$	0.029 936	0.025	
	$8s\,^3S_1 - 5p\,^3P^0_2$	0.049 789	0.042	
	$4p\,^3P^0_1 - 4d\,^3D_1$	0.453 653	0.47	
	$4p\,^3P^0_1 - 4d\,^3D_2$	0.816 559	0.84	
	$4p\,^3P^0_1 - 5d\,^3D_1$	0.128 465	0.12	
	$4p\,^3P^0_1 - 5d\,^3D_2$	0.231 216	0.21	
	$4p\,^3P^0_1 - 6d\,^3D_1$	0.058 039	0.046	

(continued)

Table 4.36 (continued)

Ions	Transition	Transition probability (10^8 s^{-1})		
		WBEPM theory	NIST*	Others
	$4p\ ^3P_1^0 - 6d\ ^3D_2$	0.104 469	0.085	
	$4d\ ^3D_1^0 - 5f\ ^3F_2^0$	0.369 943	0.42	
	$4d\ ^3D_1^0 - 6f\ ^3F_2^0$	0.172 432	0.20	
	$4d\ ^3D_2 - 5f\ ^3F_2^0$	0.068 506	0.078	
	$4d\ ^3D_2 - 6f\ ^3F_2^0$	0.390 927	0.44	
	$4d\ ^3D_2 - 6f^3F_2^0$	0.031 934	0.036	
	$4d\ ^3D_2 - 6f^3F_3^0$	0.184 05	0.22	
	$4d\ ^3D_3 - 5f^3F_2^0$	0.001 957	0.002 2	
	$4d\ ^3D_3 - 5f^3F_3^0$	0.048 864	0.055	
	$4d\ ^3D_3 - 5f^3F_4^0$	0.438 987	0.50	
	$4d\ ^3D_3 - 6f^3F_2^0$	0.000 912	0.001 0	
	$4d\ ^3D_3 - 6f^3F_3^0$	0.023 008	0.026	
	$4d\ ^3D_3 - 6f^3F_4^0$	0.209 339	0.24	
Si III[②]	$3s4p\ ^3P_2^0 - 3s4d\ ^3D_1$	0.090 753	0.095	0.088 795[a]
	$3s4p\ ^3P_2^0 - 3s4d\ ^3D_2$	0.816 816	0.88	0.799 89[a]
	$3s4p\ ^3P_2^0 - 3s4d\ ^3D_3$	3.267 6	3.4	3.202 1[a]
	$3s5p\ ^3P_0^0 - 3s4d\ ^3D_3$	0.666 34	0.63	
	$3s5p\ ^3P_0^0 - 3s5d\ ^3D_1$	0.491 45	0.51	
	$3s5p\ ^3P_1^0 - 3s4d\ ^3D_1$	0.166 57	0.16	
	$3s5p\ ^3P_1^0 - 3s4d\ ^3D_2$	0.499 68	0.49	
	$3s5p\ ^3P_2^0 - 3s4d\ ^3D_3$	0.006 658	0.006 5	
	$3s5p\ ^3P_2^0 - 3s4d\ ^3D_2$	0.099 869	0.097	
	$3s5p\ ^3P_2^0 - 3s4d\ ^3D_3$	0.559 19	0.54	
	$3s5p\ ^3P_2^0 - 3s5d\ ^3D_1$	0.024 596	0.026	
	$3s5p\ ^3P_2^0 - 3s5d\ ^3D_2$	0.221 48	0.23	
	$3s5p\ ^3P_2^0 - 3s5d^3D_3$	0.886 56	0.91	
	$3s6p\ ^3P_0^0 - 3s4d\ ^3D_3$	0.221 35	0.22	
	$3s6p\ ^3P_1^0 - 3s4d\ ^3D_1$	0.055 594	0.055	
	$3s6p\ ^3P_1^0 - 3s4d\ ^3D_2$	0.166 77	0.17	
	$3s6p\ ^3P_2^0 - 3s4d\ ^3D_3$	0.002 234	0.002 3	
	$3s6p\ ^3P_2^0 - 3s4d\ ^3D_2$	0.033 513	0.033	
	$3s6p\ ^3P_2^0 - 3s4d\ ^3D_3$	0.187 64	0.19	
	$3s7p\ ^3P_0^0 - 3s5d\ ^3D_1$	0.101 04	0.10	
	$3s7p\ ^3P_1^0 - 3s5d\ ^3D_1$	0.025 335	0.025	
	$3s7p\ ^3P_1^0 - 3s5d\ ^3D_2$	0.075 924	0.077	

(continued)

4.3 Calculation of Oscillator Strength, Transition Probability … 137

Table 4.36 (continued)

Ions	Transition	Transition probability (10^8 s^{-1})		
		WBEPM theory	NIST*	Others
	3s7p $^3P_2^0$–3s5d $^3D_1$	0.001 020	0.001 0	
	3s7p $^3P_2^0$–3s5d $^3D_2$	0.015 279	0.015	
	3s7p $^3P_2^0$–3s5d $^3D_3$	0.085 429	0.086	
	3s4d $^3D_1$–3s5f $^3F_2^0$	1.449 4	1.5	
	3s4d $^3D_1$–3s6f $^3F_2^0$	1.013 9	1.0	
	3s4d $^3D_2$–3s5f $^3F_2^0$	0.268 43	0.28	
	3s4d $^3D_2$–3s5f $^3F_3^0$	1.533 6	1.6	
	3s4d $^3D_2$–3s6f $^3F_2^0$	0.187 76	0.19	
	3s4d $^3D_3$–3s6f $^3F_4^0$	1.207 0	1.2	
	3s5d $^3D_1$–3s4f $^3F_2^0$	0.207 96	0.20	
	3s5d $^3D_1$–3s6f $^3F_2^0$	0.307 74	0.32	
	3s5d $^3D_2$–3s4f $^3F_2^0$	0.023 096	0.022	
	3s5d $^3D_2$–3s4f $^3F_3^0$	0.185 13	0.18	
	3s5d $^3D_2$–3s6f $^3F_2^0$	0.057 033	0.060	
	3s5d $^3D_2$–3s6f $^3F_3^0$	0.325 83	0.34	
	3s5d $^3D_3$–3s4f $^3F_2^0$	0.000 471	0.000 45	
	3s5d $^3D_3$–3s4f $^3F_3^0$	0.016 517	0.016	
	3s5d $^3D_3$–3s4f $^3F_4^0$	0.191 67	0.19	
	3s5d $^3D_3$–3s6f $^3F_2^0$	0.001 631	0.001 8	
	3s5d $^3D_3$–3s6f $^3F_3^0$	0.040777	0.043	
	3s5d $^3D_3$–3s6f $^3F_4^0$	0.366 896	0.39	
P IV[③]	3s4s $^3S_1$–3s4p $^3P_0^0$	2.268	2.1	2.073[b]
	3s4s $^3S_1$–3s4p $^3P_1^0$	2.281	2.09	2.073[b]
	3s4s $^3S_1$–3s4p $^3P_2^0$	2.315	2.13	2.117[b]
	3s3p $^3P_0^0$–3s3d $^3D_1$	24.92	26.3	
	3s3p $^3P_0^0$–3s4d $^3D_1$	0.466 1	0.42	
	3s3p $^3P_1^0$–3s3d $^3D_1$	18.61	19.5	
	3s3p $^3P_1^0$–3s3d $^3D_2$	33.50	35.2	
	3s3p $^3P_1^0$–3s4d $^3D_1$	0.339 9	0.32	
	3s3p $^3P_1^0$–3s4d $^3D_2$	0.613 5	0.55	
	3s3p $^3P_2^0$–3s3d $^3D_1$	1.230	1.3	
	3s3p $^3P_2^0$–3s3d $^3D_2$	11.07	11.6	
	3s3p $^3P_2^0$–3s3d $^3D_3$	44.27	46.4	
	3s3p $^3P_2^0$–3s4d $^3D_1$	0.021 36	0.021	
	3s3p $^3P_2^0$–3s4d $^3D_2$	0.192 8	0.19	

(continued)

Table 4.36 (continued)

Ions	Transition	Transition probability (10^8 s^{-1})		
		WBEPM theory	NIST*	Others
	3s3p $^3P_2^0$–3s4d $^3D_3$	0.774 7	0.75	
	3s3d $^3D_1$–3s4p $^3P_0^0$	6.807		6.736[b]
	3s3d $^3D_1$–3s4p $^3P_1^0$	1.703		1.676[b]
	3s3d $^3D_1$–3s4p $^3P_2^0$	0.068 17		0.067 76[b]
	3s3d $^3D_2$–3s4p $^3P_1^0$	5.108		5.29[b]
	3s3d $^3D_2$–3s4p $^3P_2^0$	1.023		1.17[b]
	3s3d $^3D_3$–3s4p $^3P_2^0$	5.726		5.692[b]
	3s4p $^3P_0^0$–3s4d $^3D_1$	3.429		
	3s4p $^3P_1^0$–3s4d $^3D_1$	2.562		
	3s4p $^3P_1^0$–3s4d $^3D_2$	4.614		
	3s4p $^3P_2^0$–3s4d $^3D_1$	0.169 2		
	3s4p $^3P_2^0$–3s4d $^3D_2$	1.524		
	3s4p $^3P_2^0$–3s4d $^3D_3$	6.98		
S V[③]	3s4s $^3S_1$–3s4p $^3P_0^0$	3.376		3.4[d]
	3s4s $^3S_1$–3s4p $^3P_1^0$	3.387		2.8[d]
	3s4s $^3S_1$–3s4p $^3P_2^0$	3.473		3.6[d]
	3s3p $^3P_0^0$–3s3d $^3D_1$	36.27	36.2	38.3[c]
	3s3p $^3P_1^0$–3s3d $^3D_1$	27.4	27.0	28.5[c]
	3s3p $^3P_1^0$–3s3d $^3D_2$	48.68	48.7	
	3s3p $^3P_2^0$–3s3d $^3D_1$	1.781	1.8	1.88[c]
	3s3p $^3P_2^0$–3s3d $^3D_2$	16.3	16.0	16.9[c]
	3s3p $^3P_2^0$–3s3d $^3D_3$	64.12	63.9	67.8[c]
	3s3d $^3D_1$–3s4p $^3P_0^0$	16.55		12.627[b], 12[d]
	3s3d $^3D_1$–3s4p $^3P_1^0$	4.136		2.608[b], 2.5[d]
	3s3d $^3D_1$–3s4p $^3P_2^0$	0.165 2		0.093[b], 0.11[d]
	3s3d $^3D_2$–3s4p $^3P_1^0$	12.41		7.545[b], 7.4[d]
	3s3d $^3D_2$–3s4p $^3P_2^0$	2.478		1.633[b], 1.8[d]
	3s3d $^3D_3$–3s4p $^3P_2^0$	13.88		11.260[b], 11[d]
	3s4p $^3P_0^0$–3s4d $^3D_1$	5.187		
	3s4p $^3P_1^0$–3s4d $^3D_1$	3.882		
	3s4p $^3P_1^0$–3s4d $^3D_2$	6.994		
	3s4p $^3P_2^0$–3s4d $^3D_1$	0.254 4		
	3s4p $^3P_2^0$–3s4d $^3D_2$	2.292		
	3s4p $^3P_2^0$–3s4d $^3D_3$	9.179		

(continued)

4.3 Calculation of Oscillator Strength, Transition Probability ...

Table 4.36 (continued)

Ions	Transition	Transition probability (10^8 s^{-1})		
		WBEPM theory	NIST*	Others
Cl VI[③]	3s4s $^3S_1$–3s4p $^3P_0^0$	4.706		4.132^b, 4.7^d
	3s4s $^3S_1$–3s4p $^3P_1^0$	4.758		4.179^b, 4.7^d
	3s4s $^3S_3$–3s4p $^3P_2^0$	4.887		4.292^b, 4.9^d
	3s3p $^3P_0^0$–3s3d $^3D_1$	47.25	46.2	
	3s3p $^3P_1^0$–3s3d $^3D_1$	35.15	34.4	
	3s3p $^3P_2^0$–3s3d $^3D_2$	63.30	61.9	
	3s3p $^3P_2^0$–3s3d $^3D_1$	2.305	2.3	
	3s3p $^3P_2^0$–3s3d $^3D_2$	20.75	20.3	
	3s3p $^3P_2^0$–3s3d $^3D_3$	83.03	81.2	
	3s3d $^3D_1$–3s4p $^3P_0^0$	32.55		30.84^b, 32^d
	3s3d $^3D_1$–3s4p $^3P_1^0$	8.127		7.670^b, 7.9^d
	3s3d $^3D_1$–3s4p $^3P_2^0$	0.324 1		0.306 7^b, 0.31^d
	3s3d $^3D_2$–3s4p $^3P_1^0$	24.38		23.15^b, 24^d
	3s3d $^3D_2$–3s4p $^3P_2^0$	4.861		4.611^b, 4.8^d
	3s3d $^3D_3$–3s4p $^3P_2^0$	27.22		26.44^b, 27^d
	3s4P $^3P_0^0$–3s4d $^3D_1$	6.785		
	3s4P $^3P_1^0$–3s4d $^3D_1$	5.049		
	3s4P $^3P_1^0$–3s4d $^3D_2$	9.102		
	3s4P $^3P_2^0$–3s4d $^3D_3$	0.330 2		
	3s4P $^3P_2^0$–3s4d $^3D_2$	2.976		
	3s4P $^3P_2^0$–3s4d $^3D_3$	11.94		
Ar VII[③]	3s3p $^3P_0^0$–3s3d $^3D_1$	58.63	56.4	59.2^c
	3s3p $^3P_1^0$–3s3d $^3D_1$	43.53	41.8	44.0^c
	3s3p $^1P_1^0$–3s3d $^3D_2$	78.40	75.3	79.2^c
	3s3p $^3P_2^0$–3s3d $^3D_1$	2.840	2.7	2.88^c
	3s3p $^3P_2^0$–3s3d $^3D_2$	25.57	24.5	26.0^c
	3s3p $^3P_2^0$–3s3d $^3D_3$	102.4	98.0	104^c

Notes [①] Zheng NW, Fan J, Ma DX, et al. J Phys Soc JPN, 2003, 72:3091–3096
[②] Fan J, Zhang TY, Zheng NW, et al. Chin J Chem Phys, 2007, 20:265–272
[③] Fan J, Zheng NW. Chem Phys Lett, 2004, 400:273–278
[In which * is taken from Fuhr JR, Martin WC, Musgrove A, et al. NIST Atomic Spectroscopic Database, Version 2.0, 1996; http://physics.nist.gov (Select "Physical Reference Data")
a. Tachiev G, Fischer CF. http://www.vuse.vanderbilt.edu/~cff/mchf.collection
b. Brage T, Hibbert A. J Phys B, 1989, 22:713
c. Christensen RB, Norcross DW, Pradhan AK. Phys Rev A, 1986, 34:4704
d. Reistad N, Brage T, Ekberg JO, et al. Phys Scr, 1984, 30:249]

Table 4.37 Transition probability between electron configurations of lithium-like ions

Ions	Transition	Transition probability ($\times 10^8$ Hz)		
		WBEPM theory	NIST[①]	Others[②]
Li I	2p–2s	0.366 66	0.372	0.366
	3p–2s	0.005 98	0.011 7	0.008 984
	4p–2s	0.008 48	0.014 2	0.011 68
	2p–3s	0.337 34	0.349	0.329 8
	3p–3s	0.037 22	0.037 7	0.037 23
	4p–3s	6.854 7E−6	3.69E−5	6.728E−6
	2p–4s	0.102 28	0.101	0.102 2
	3p–4s	0.074 63	0.074 6	0.074 31
	4p–4s	0.007 73	0.007 72	0.007 737
	2p–5s	0.046 3	0.046	0.046 53
	3p–5s	0.027 87	0.027 6	0.028 12
	4p–5s	0.022 5	0.022 5	0.022 45
	5p–5s	0.002 34		0.002 34
	3d–2p	0.691 39	0.716	0.682 2
	4d–2p	0.235 89	0.230	0.231 6
	5d–2p	0.110 52	0.106	0.108 3
	6d–2p	0.061 09		0.059 91
	3d–3p	3.814 28E−5	3.81E−5	3.789E−5
	4d–3p	0.067 63	0.068 5	0.067 68
	3d–4p	0.005 5	0.005 2	0.005 419
	4d–4p	1.286 96E−5	1.28E−5	1.278E−5
	5d–4p	0.013 46	0.013 6	0.013 51
	3d–5p	0.002 29	0.002 31	0.002 316
	4d–5p	0.002 83	0.002 86	0.002 808
	5d–5p	4.818 63E−6	4.78E−6	4.78E−6
	4d–6p	0.001 38	0.001 39	0.001 378
	5d–6p	0.001 4	0.001 42	0.001 372
Be II	2p–2s	1.134 53	1.146	1.119
	2p–3s	4.068 97	2.9	3.982
	4p–3s	0.133 81	0.19	0.140 4
	5p–3s	0.099 2	0.142	0.105 6
	2p–4s	1.390 74	0.94	1.387
	3p–4s	0.964 67	0.66	0.963 8
	5p–4s	0.024 84	0.030	0.025 53
	6p–4s	0.021 53	0.025 6	0.022 36

(continued)

4.3 Calculation of Oscillator Strength, Transition Probability ...

Table 4.37 (continued)

Ions	Transition	Transition probability ($\times 10^8$ Hz)		
		WBEPM theory	NIST[1]	Others[2]
	3p–5s	0.407 17	0.28	0.411 4
	4p–6s	0.144 96	0.102	0.145 8
	3d–2p	11.149 8	11	10.98
	4d–3p	1.077 17	1.1	1.079
	5d–3p	0.559 61	0.54	0.555
	6d–4p	0.133 2	0.132	0.132 6
B III	2p–2s	1.902 63		1.876
	3p–2s	11.715 5		12.36
	4p–2s	5.920 6		6.359
	3p–4s	3.942 93		3.948
	4p–4s	0.049 1		0.049 01
C IV	2p–2s	2.648 41	2.64	2.620
	3p–2s	43.511 62	44.9	45.19
	3p–3s	0.314 53	0.316	0.315 2
	3d–2p	176.042 28	180	173.9
N V	2p–2s	3.380 75	3.38	3.365
	3p–2s	115.199 33	118	118.6
	3p–3s	0.409 18	0.408	0.412 0
	3d–2p	427.081 7	430	422.4
O VI	2p–2s	4.105 89	4.08	4.122
	3p–2s	250.693 62	254	256.4
	3p–3s	0.503 35	0.506	0.511 1
F VII	2p–2s	4.827 81		4.900
	3p–2s	479.291 99		487.7
	4p–2s	220.484 18		227.3
	5p–2s	115.398 71		119.3
	3d–4p	11.065 52		10.98
	4d–4p	0.005 5		0.005 616
	5d–4p	33.896 85		33.95
	6d–4p	20.478 7		20.40

Data source Zheng NW, Sun YJ, Wang T, et al. Int J Quantum Chem, 2000, 76:51
Notes [1] Weast RC. CRC handbook of chemistry and physics: a ready-reference book of chemical and physical data, 70th edn. CRC Press, Inc., Boca Raton, Florida, 189–1990
[2] Lindgard A, Nielsen SE. At Data Nucl Data Tab, 1997, 19:533

Table 4.38 Transition probability and oscillator strength of resonance transition of multiplet nitrogen atom

Transition	Transition probability of WBEPM theory (10^8 s^{-1})	Oscillator strength		
		WBEPM theory	Other theories	Experimental value
$2p^3\,^4S^0 - 2p^23s\,^4P$	3.302	0.213 9	0.283 9[a], 0.241[b] 0.241[c], 0.288[d] 0.231[e], 0.262[f]	0.266[h], 0.275[i]
$2p^3\,^4S^0 - 2p^24s\,^4P$	0.800 5	0.033 49	0.025 3[a], 0.034[b] 0.033[c], 0.025 8[f]	0.030[h], 0.027[j]
$2p^3\,^4S^0 - 2p^25s\,^4P$	0.330 2	0.012 30	0.010 9[f]	
$2p^3\,^4S^0 - 2p^26s\,^4P$	0.170 2	0.006 034	0.005 35[f]	
$2p^3\,^4S^0 - 2p^23d\,^4P$	2.031	0.083 10	0.076 1[a], 0.065[b] 0.062[c], 0.12[d] 0.078[e], 0.0781[f]	0.067[h], 0.075[j]
$2p^3\,^4S^0 - 2p^24d\,^4P$	1.135	0.041 95	0.036 9[f]	
$2p^3\,^4S^0 - 2p^25d\,^4P$	0.664 8	0.023 49	0.019 7[f]	
$2p^3\,^4S^0 - 2p^26d\,^4P$	0.4276	0.014 75	0.011 7[f]	
$2p^3\,^2D^0 - 2p^23s\,^2P$	3.270	0.065 60	0.102[d], 0.112[g]	0.071[i]
$2p^3\,^2D^0 - 2p^24s\,^2P$	0.941 0	0.011 73	0.032 2[d], 0.013 8[g]	0.013[g]
$2p^3\,^2D^0 - 2p^23d\,^2P$	0.038 10	0.000 47	0.000 45[d]	
$2p^3\,^2D^0 - 2p^23d\,^2F$	1.235	0.035 37	0.047[d], 0.035[g]	0.032[j]
$2p^3\,^2D^0 - 2p^23d\,^2D$	0.268 5	0.005 455	0.008 1[d], 0.009 5[g]	
$2p^3\,^2P^0 - 2p^23s\,^2P$	1.358	0.061 92	0.09[d], 0.092[g]	0.061[i]

(continued)

out that WBEPM Theory can be widely used for the study of transition properties for lower excited states, higher excited states (especially higher excited states), simple systems and complicated systems with accurate results, and the calculation process is much simpler than other theoretical methods, so it is a very good theoretical method [144, 145, 149] (Table 4.43).

Table 4.38 (continued)

Transition	Transition probability of WBEPM theory (10^8 s^{-1})	Oscillator strength		
		WBEPM theory	Other theories	Experimental value
$2p^3\,^2P^0 - 2p^24s\,^2P$	0.4041	0.01067	0.013^d, 0.005^g	

Data source Zheng NW, Wang T. Chem Phys, 2002, 282:31–36

Notes a. Tong M, Fischer CF, Sturesson L. J Phys B, 1994, 27:4819 (MCHF method)
b. and c. Robinson DJR, Hibbert A. J Phys B, 1997, 30:4813 (CI method)
d. Aashamar K, Luke TM, Talman JD. Phys Scr, 1983, 27:267 (MCOPM method)
e. Beck DR, Nicolaides CA. J Quant Spectrosc Radiat Transfer, 1976, 16:297 (MCHF method)
f. Bell KL, Berrington KA. J Phys B, 1991, 24:933 (R-matrix method)
g. Wiess WL, Smith MW, Glennon BM. Atomic transition probabilities: Natl. Bur. Stand. (US) Pub. NSRDSNBS 4. Washington, 1996
h. Lugger PM, York DG, Blanchard T, et al. Astrophys J, 1978, 224:1059 (Experimental value)
i. Goldbach C, Martin M, Nollez G, et al. Astron Astrophys, 1986, 161:47 (Experimental value)
j. Goldbach C, Ludtke T, Martin M, et al. Astron Astrophys, 1992, 266:605 (Experimental value)

Table 4.39 Radiative lifetime of some excited states of carbon atom (ns)

Excited state	WBEPM theory	Other methods[①]	Experimental value[②]
$2p3s\,^3P^0_0$	2.676	2.910	
$2p3s\,^3P^0_1$	2.682	2.908	
$2p3s\,^3P^0_2$	2.694	2.904	
$2p3s\,^3P^0_1$	2.704	2.657	2.7 ± 0.2
$2p3p\,^3D_1$	55.42	53.33	
$2p3p\,^3D_2$	55.33	53.28	
$2p3p\,^3D_3$	55.18	53.25	
$2p3p\,^3P_0$	35.28	32.40	
$2p3p\,^3P_1$	35.25	32.40	
$2p3p\,^3P_2$	35.22	32.38	
$2p3p\,^3S_1$	42.57	41.13	
$2p3p\,^1D_2$	36.24	33.60	
$2p3p\,^1P_1$	127.3	114.3	
$2p3p\,^1S_0$	28.39	27.71	
$2p4s\,^3P^0_0$	9.066	9.505	
$2p4s\,^3P^0_1$	9.099	9.260	
$2p4s\,^3P^0_2$	9.354	8.726	
$2p4p\,^3P_0$	152.2	133.1	
$2p4p\,^3P_1$	150.2	133.6	
$2p4p\,^3P_2$	148.0	133.0	

(continued)

Table 4.39 (continued)

Excited state	WBEPM theory	Other methods[1]	Experimental value[2]
$2p4p\ ^3S_1$	236.4	201.5	
$2p4p\ ^1D_2$	116.9	114.5	121 ± 6
$2p4p\ ^1P_1$	274.2	226.2	
$2p4p\ ^1S_0$	77.22	72.82	81 ± 4
$2p3d\ ^3F_4^0$	37.77	36.44	
$2p3d\ ^3D_3^0$	4.524	4.240	
$2p3d\ ^3D_1^0$	4.476	4.328	
$2p3d\ ^1F_3^0$	5.097	5.283	

Data source Zheng NW, Wang T. Astrophys J Suppl Ser, 2002, 143:231

Notes [1] Hibbert A, Biement E, Godefroid M, et al. Astron Astrophys Suppl Ser, 1993, 99:179 (CI method)
[2] O'Brain TR, Lawler JE. Quant J Spectrosc Radiat Transfer, 1997, 57:309

Table 4.40 Transition probability ($10^8\ s^{-1}$) between energy levels for copper, silver and gold atoms[1]

Atom	Transition	WBEPM theory	Transition	WBEPM theory
Cu I	$4s\,^2S_{1/2} - 4p\,^2P_{3/2}^0$	1.542e+00	$4s\,^2S_{1/2} - 5p\,^2P_{3/2}^0$	6.061e−02
	$4s\,^2S_{1/2} - 6p\,^2P_{3/2}^0$	7.401e−04	$4s\,^2S_{1/2} - 7p\,^2P_{3/2}^0$	2.724e−02
	$4s\,^2S_{1/2} - 8p\,^2P_{3/2}^0$	1.547e−02	$4s\,^2S_{1/2} - 9p\,^2P_{3/2}^0$	8.953e−03
	$4s\,^2S_{1/2} - 10p\,^2P_{3/2}^0$	5.948e−03	$4s\,^2S_{1/2} - 11p\,^2P_{3/2}^0$	4.215e−03
	$4s\,^2S_{1/2} - 12p\,^2P_{3/2}^0$	3.119e−03	$4s\,^2S_{1/2} - 13p\,^2P_{3/2}^0$	2.389e−03
	$4s\,^2S_{1/2} - 14p\,^2P_{3/2}^0$	1.892e−03	$4s\,^2S_{1/2} - 15p\,^2P_{3/2}^0$	1.504e−03
	$5s\,^2S_{1/2} - 5p\,^2P_{3/2}^0$	1.147e−01	$5s\,^2S_{1/2} - 6p\,^2P_{3/2}^0$	2.438e−05
	$5s\,^2S_{1/2} - 7p\,^2P_{3/2}^0$	6.191e−03	$5s\,^2S_{1/2} - 8p\,^2P_{3/2}^0$	3.143e−03
	$5s\,^2S_{1/2} - 9p\,^2P_{3/2}^0$	1.656e−03	$5s\,^2S_{1/2} - 10p\,^2P_{3/2}^0$	1.061e−03
	$5s\,^2S_{1/2} - 11p\,^2P_{3/2}^0$	7.406e−04	$5s\,^2S_{1/2} - 12p\,^2P_{3/2}^0$	5.450e−04
	$5s\,^2S_{1/2} - 13p\,^2P_{3/2}^0$	4.179e−04	$5s\,^2S_{1/2} - 14p\,^2P_{3/2}^0$	3.339e−04
	$5s\,^2S_{1/2} - 15p\,^2P_{3/2}^0$	2.656e−04	$6s\,^2S_{1/2} - 6p\,^2P_{3/2}^0$	1.417e−02
	$6s\,^2S_{1/2} - 7p\,^2P_{3/2}^0$	4.351e−03	$6s\,^2S_{1/2} - 8p\,^2P_{3/2}^0$	1.759e−03
	$6s\,^2S_{1/2} - 9p\,^2P_{3/2}^0$	8.437e−04	$6s\,^2S_{1/2} - 10p\,^2P_{3/2}^0$	5.118e−04
	$6s\,^2S_{1/2} - 11p\,^2P_{3/2}^0$	3.456e−04	$6s\,^2S_{1/2} - 12p\,^2P_{3/2}^0$	2.490e−04
	$6s\,^2S_{1/2} - 13p\,^2P_{3/2}^0$	1.882e−04	$6s\,^2S_{1/2} - 14p\,^2P_{3/2}^0$	1.489e−04
	$6s\,^2S_{1/2} - 15p\,^2P_{3/2}^0$	1.176e−04	$7s\,^2S_{1/2} - 7p\,^2P_{3/2}^0$	7.827e−03
	$7s\,^2S_{1/2} - 8p\,^2P_{3/2}^0$	1.532e−03	$7s\,^2S_{1/2} - 9p\,^2P_{3/2}^0$	6.134e−04

(continued)

4.3 Calculation of Oscillator Strength, Transition Probability ...

Table 4.40 (continued)

Atom	Transition	WBEPM theory	Transition	WBEPM theory
	$7s\,^2S_{1/2} - 10p\,^2P^0_{3/2}$	3.413e−04	$7s\,^2S_{1/2} - 11p\,^2P^0_{3/2}$	2.198e−04
	$7s\,^2S_{1/2} - 12p\,^2P^0_{3/2}$	1.538e−04	$7s\,^2S_{1/2} - 13p\,^2P^0_{3/2}$	1.140e−04
	$7s\,^2S_{1/2} - 14p\,^2P^0_{3/2}$	8.892e−05	$7s\,^2S_{1/2} - 15p\,^2P^0_{3/2}$	6.955e−05
	$8s\,^2S_{1/2} - 8p\,^2P^0_{3/2}$	2.845e−03	$8s\,^2S_{1/2} - 9p\,^2P^0_{3/2}$	5.977e−04
	$8s\,^2S_{1/2} - 10p\,^2P^0_{3/2}$	2.775e−04	$8s\,^2S_{1/2} - 11p\,^2P^0_{3/2}$	1.650e−04
	$8s\,^2S_{1/2} - 12p\,^2P^0_{3/2}$	1.106e−04	$8s\,^2S_{1/2} - 13p\,^2P^0_{3/2}$	7.986e−05
	$8s\,^2S_{1/2} - 14p\,^2P^0_{3/2}$	6.119e−05	$8s\,^2S_{1/2} - 15p\,^2P^0_{3/2}$	4.729e−05
	$9s\,^2S_{1/2} - 9p\,^2P^0_{3/2}$	1.191e−03	$9s\,^2S_{1/2} - 10p\,^2P^0_{3/2}$	2.856e−04
	$9s\,^2S_{1/2} - 11p\,^2P^0_{3/2}$	1.432e−04	$9s\,^2S_{1/2} - 12p\,^2P^0_{3/2}$	8.915e−05
	$9s\,^2S_{1/2} - 13p\,^2P^0_{3/2}$	6.184e−05	$9s\,^2S_{1/2} - 14p\,^2P^0_{3/2}$	4.621e−05
	$9s\,^2S_{1/2} - 15p\,^2P^0_{3/2}$	3.516e−05	$4p\,^2P^0_{3/2} - 4d\,^2D_{5/2}$	3.480e−01
	$4p\,^2P^0_{3/2} - 5d\,^2D_{5/2}$	8.236e−02	$4p\,^2P^0_{3/2} - 6d\,^2D_{5/2}$	3.338e−02
	$4p\,^2P^0_{3/2} - 7d\,^2D_{5/2}$	1.683e−02	$4p\,^2P^0_{3/2} - 8d\,^2D_{5/2}$	9.631e−03
	$4p\,^2P^0_{3/2} - 9d\,^2D_{5/2}$	5.998e−03	$4p\,^2P^0_{3/2} - 10d\,^2D_{5/2}$	3.969e−03
	$4p\,^2P^0_{3/2} - 11d\,^2D_{5/2}$	2.782e−03	$5p\,^2P^0_{3/2} - 4d\,^2D_{5/2}$	1.574e−04
	$5p\,^2P^0_{3/2} - 5d\,^2D_{5/2}$	5.112e−02	$5p\,^2P^0_{3/2} - 6d\,^2D_{5/2}$	1.847e−02
	$5p\,^2P^0_{3/2} - 7d\,^2D_{5/2}$	9.012e−03	$5p\,^2P^0_{3/2} - 8d\,^2D_{5/2}$	5.113e−03
	$5p\,^2P^0_{3/2} - 9d\,^2D_{5/2}$	3.183e−03	$5p\,^2P^0_{3/2} - 10d\,^2D_{5/2}$	2.112e−03
	$5p\,^2P^0_{3/2} - 11d\,^2D_{5/2}$	1.485e−03	$6p\,^2P^0_{3/2} - 5d\,^2D_{5/2}$	7.826e−04
	$6p\,^2P^0_{3/2} - 6d\,^2D_{5/2}$	7.421e−03	$6p\,^2P^0_{3/2} - 7d\,^2D_{5/2}$	3.710e−03
	$6p\,^2P^0_{3/2} - 8d\,^2D_{5/2}$	2.048e−03	$6p\,^2P^0_{3/2} - 9d\,^2D_{5/2}$	1.36e−03
	$6p\,^2P^0_{3/2} - 10d\,^2D_{5/2}$	7.954e−04	$6p\,^2P^0_{3/2} - 11d\,^2D_{5/2}$	5.492e−04
	$7p\,^2P^0_{3/2} - 6d\,^2D_{5/2}$	2.157c−06	$7p\,^2P^0_{3/2} - 7d\,^2D_{5/2}$	4.675e−03
	$7p\,^2P^0_{3/2} - 8d\,^2D_{5/2}$	2.345e−03	$7p\,^2P^0_{3/2} - 9d\,^2D_{5/2}$	1.397e−03
	$7p\,^2P^0_{3/2} - 10d\,^2D_{5/2}$	9.130e−04	$7p\,^2P^0_{3/2} - 11d\,^2D_{5/2}$	6.363e−04
	$8p\,^2P^0_{3/2} - 7d\,^2D_{5/2}$	4.424e−07	$8p\,^2P^0_{3/2} - 8d\,^2D_{5/2}$	1.813e−03
	$8p\,^2P^0_{3/2} - 9d\,^2D_{5/2}$	1.017e−03	$8p\,^2P^0_{3/2} - 10d\,^2D_{5/2}$	6.452e−04
	$8p\,^2P^0_{3/2} - 11d\,^2D_{5/2}$	4.425e−04	$9p\,^2P^0_{3/2} - 8d\,^2D_{5/2}$	3.407e−09
	$9p\,^2P^0_{3/2} - 9d\,^2D_{5/2}$	7.763e−04	$9p\,^2P^0_{3/2} - 10d\,^2D_{5/2}$	4.779e−04
	$9p\,^2P^0_{3/2} - 11d\,^2D_{5/2}$	3.19e−04	$10p\,^2P^0_{3/2} - 9d\,^2D_{5/2}$	8.334e−16
	$10p\,^2P^0_{3/2} - 10d\,^2D_{5/2}$	3.792e−04	$10p\,^2P^0_{3/2} - 11d\,^2D_{5/2}$	2.513e−04
	$11p\,^2P^0_{3/2} - 10d\,^2D_{5/2}$	1.129e−10	$11p\,^2P^0_{3/2} - 11d\,^2D_{5/2}$	2.056e−04

(continued)

Table 4.40 (continued)

Atom	Transition	WBEPM theory	Transition	WBEPM theory
	$12p\,^2P^0_{3/2} - 11d\,^2D_{5/2}$	1.691e−11	$4d\,^2D_{5/2} - 4f\,^2F^0_{7/2}$	1.408e−01
	$4d\,^2D_{5/2} - 5f\,^2F^0_{7/2}$	5.502e−02	$5d\,^2D_{5/2} - 4f\,^2F^0_{7/2}$	9.175e−08
	$5d\,^2D_{5/2} - 5f\,^2F^0_{7/2}$	2.481e−02	$6d\,^2D_{5/2} - 5f\,^2F^0_{7/2}$	1.938e−08
	$4s\,^2S_{1/2} - 4p\,^2P^0_{1/2}$	1.512e+00	$4s\,^2S_{1/2} - 5p\,^2P^0_{1/2}$	6.064e−02
	$4s\,^2S_{1/2} - 6p\,^2P^0_{1/2}$	1.100e−02	$4s\,^2S_{1/2} - 7p\,^2P^0_{1/2}$	2.490e−03
	$4s\,^2S_{1/2} - 8p\,^2P^0_{1/2}$	1.982e−02	$4s\,^2S_{1/2} - 9p\,^2P^0_{1/2}$	1.031e−02
	$4s\,^2S_{1/2} - 10p\,^2P^0_{1/2}$	6.510e−03	$4s\,^2S_{1/2} - 11p\,^2P^0_{1/2}$	4.496e−03
	$4s\,^2S_{1/2} - 12p\,^2P^0_{1/2}$	3.275e−03	$4s\,^2S_{1/2} - 13p\,^2P^0_{1/2}$	2.457e−03
	$5s\,^2S_{1/2} - 5p\,^2P^0_{1/2}$	1.148e−01	$5s\,^2S_{1/2} - 6p\,^2P^0_{1/2}$	2.527e−03
	$5s\,^2S_{1/2} - 7p\,^2P^0_{1/2}$	2.097e−03	$5s\,^2S_{1/2} - 8p\,^2P^0_{1/2}$	4.504e−03
	$5s\,^2S_{1/2} - 9p\,^2P^0_{1/2}$	2.043e−03	$5s\,^2S_{1/2} - 10p\,^2P^0_{1/2}$	1.216e−03
	$5s\,^2S_{1/2} - 11p\,^2P^0_{1/2}$	8.171e−04	$5s\,^2S_{1/2} - 12p\,^2P^0_{1/2}$	5.873e−04
	$5s\,^2S_{1/2} - 13p\,^2P^0_{1/2}$	4.363e−04	$6s\,^2S_{1/2} - 6p\,^2P^0_{1/2}$	1.887e−02
	$6s\,^2S_{1/2} - 7p\,^2P^0_{1/2}$	4.793e−04	$6s\,^2S_{1/2} - 8p\,^2P^0_{1/2}$	2.455e−03
	$6s\,^2S_{1/2} - 9p\,^2P^0_{1/2}$	1.034e−03	$6s\,^2S_{1/2} - 10p\,^2P^0_{1/2}$	5.858e−04
	$6s\,^2S_{1/2} - 11p\,^2P^0_{1/2}$	3.814e−04	$6s\,^2S_{1/2} - 12p\,^2P^0_{1/2}$	2.685e−04
	$6s\,^2S_{1/2} - 13p\,^2P^0_{1/2}$	1.966e−04	$7s\,^2S_{1/2} - 7p\,^2P^0_{1/2}$	2.349e−03
	$7s\,^2S_{1/2} - 8p\,^2P^0_{1/2}$	1.996e−03	$7s\,^2S_{1/2} - 9p\,^2P^0_{1/2}$	7.352e−04
	$7s\,^2S_{1/2} - 10p\,^2P^0_{1/2}$	3.868e−04	$7s\,^2S_{1/2} - 11p\,^2P^0_{1/2}$	2.412e−04
	$7s\,^2S_{1/2} - 12p\,^2P^0_{1/2}$	1.653e−04	$7s\,^2S_{1/2} - 13p\,^2P^0_{1/2}$	1.189e−04
	$8s\,^2S_{1/2} - 8p\,^2P^0_{1/2}$	3.071e−03	$8s\,^2S_{1/2} - 9p\,^2P^0_{1/2}$	6.912e−04
	$8s\,^2S_{1/2} - 10p\,^2P^0_{1/2}$	3.105e−04	$8s\,^2S_{1/2} - 11p\,^2P^0_{1/2}$	1.800e−04
	$8s\,^2S_{1/2} - 12p\,^2P^0_{1/2}$	1.185e−04	$8s\,^2S_{1/2} - 13p\,^2P^0_{1/2}$	8.315e−05
	$9s\,^2S_{1/2} - 9p\,^2P^0_{1/2}$	1.248e−03	$9s\,^2S_{1/2} - 10p\,^2P^0_{1/2}$	3.129e−04
	$9s\,^2S_{1/2} - 11p\,^2P^0_{1/2}$	1.548e−04	$9s\,^2S_{1/2} - 12p\,^2P^0_{1/2}$	9.502e−05
	$9s\,^2S_{1/2} - 13p\,^2P^0_{1/2}$	6.425e−05		
Ag I	$5s\,^2S_{1/2} - 5p\,^2P^0_{3/2}$	1.701e+00	$5s\,^2S_{1/2} - 6p\,^2P^0_{3/2}$	8.520e−01
	$5s\,^2S_{1/2} - 7p\,^2P^0_{3/2}$	1.884e−02	$5s\,^2S_{1/2} - 8p\,^2P^0_{3/2}$	1.040e−04
	$5s\,^2S_{1/2} - 9p\,^2P^0_{3/2}$	2.283e−07	$5s\,^2S_{1/2} - 10p\,^2P^0_{3/2}$	2.430e−10
	$6s\,^2S_{1/2} - 6p\,^2P^0_{3/2}$	1.256e−01	$6s\,^2S_{1/2} - 7p\,^2P^0_{3/2}$	2.729e−01
	$6s\,^2S_{1/2} - 8p\,^2P^0_{3/2}$	3.285e−02	$6s\,^2S_{1/2} - 9p\,^2P^0_{3/2}$	1.030e−03
	$6s\,^2S_{1/2} - 10p\,^2P^0_{3/2}$	1.286e−05	$7s\,^2S_{1/2} - 7p\,^2P^0_{3/2}$	2.605e−02

(continued)

4.3 Calculation of Oscillator Strength, Transition Probability ...

Table 4.40 (continued)

Atom	Transition	WBEPM theory	Transition	WBEPM theory
	$7s\,^2S_{1/2} - 8p\,^2P^0_{3/2}$	1.017e−01	$7s\,^2S_{1/2} - 9p\,^2P^0_{3/2}$	2.676e−02
	$7s\,^2S_{1/2} - 10p\,^2P^0_{3/2}$	1.975e−03	$8s\,^2S_{1/2} - 8p\,^2P^0_{3/2}$	8.019e−03
	$8s\,^2S_{1/2} - 9p\,^2P^0_{3/2}$	4.458e−02	$8s\,^2S_{1/2} - 10p\,^2P^0_{3/2}$	1.903e−02
	$9s\,^2S_{1/2} - 9p\,^2P^0_{3/2}$	3.097e−03	$9s\,^2S_{1/2} - 10p\,^2P^0_{3/2}$	2.198e−02
	$10s\,^2S_{1/2} - 10p\,^2P^0_{3/2}$	1.418e−03	$5p\,^2P^0_{3/2} - 5d\,^2D_{5/2}$	7.568e−01
	$5p\,^2P^0_{3/2} - 6d\,^2D_{5/2}$	1.185e−01	$5p\,^2P^0_{3/2} - 7d\,^2D_{5/2}$	2.714e−03
	$5p\,^2P^0_{3/2} - 8d\,^2D_{5/2}$	2.066e−05	$5p\,^2P^0_{3/2} - 9d\,^2D_{5/2}$	7.214e−08
	$5p\,^2P^0_{3/2} - 10d\,^2D_{5/2}$	1.370e−10	$5p\,^2P^0_{3/2} - 11d\,^2D_{5/2}$	1.566e−13
	$5p\,^2P^0_{3/2} - 12d\,^2D_{5/2}$	1.241e−16	$6p\,^2P^0_{3/2} - 5d\,^2D_{5/2}$	1.772e−05
	$6p\,^2P^0_{3/2} - 6d\,^2D_{5/2}$	1.863e−01	$6p\,^2P^0_{3/2} - 7d\,^2D_{5/2}$	9.384e−02
	$6p\,^2P^0_{3/2} - 8d\,^2D_{5/2}$	7.898e−03	$6p\,^2P^0_{3/2} - 9d\,^2D_{5/2}$	2.299e−04
	$6p\,^2P^0_{3/2} - 10d\,^2D_{5/2}$	3.108e−06	$6p\,^2P^0_{3/2} - 11d\,^2D_{5/2}$	2.287e−08
	$6p\,^2P^0_{3/2} - 12d\,^2D_{5/2}$	1.069e−10	$7p\,^2P^0_{3/2} - 6d\,^2D_{5/2}$	2.970e−06
	$7p\,^2P^0_{3/2} - 7d\,^2D_{5/2}$	6.162e−02	$7p\,^2P^0_{3/2} - 8d\,^2D_{5/2}$	5.449e−02
	$7p\,^2P^0_{3/2} - 9d\,^2D_{5/2}$	9.167e−03	$7p\,^2P^0_{3/2} - 10d\,^2D_{5/2}$	5.690e−04
	$7p\,^2P^0_{3/2} - 11d\,^2D_{5/2}$	1.697e−05	$7p\,^2P^0_{3/2} - 12d\,^2D_{5/2}$	2.936e−07
	$8p\,^2P^0_{3/2} - 7d\,^2D_{5/2}$	9.069e−07	$8p\,^2P^0_{3/2} - 8d\,^2D_{5/2}$	2.496e−02
	$8p\,^2P^0_{3/2} - 9d\,^2D_{5/2}$	3.150e−02	$8p\,^2P^0_{3/2} - 10d\,^2D_{5/2}$	8.364e−03
	$8p\,^2P^0_{3/2} - 11d\,^2D_{5/2}$	8.671e−04	$8p\,^2P^0_{3/2} - 12d\,^2D_{5/2}$	4.622e−05
	$9p\,^2P^0_{3/2} - 8d\,^2D_{5/2}$	3.959e−07	$9p\,^2P^0_{3/2} - 9d\,^2D_{5/2}$	1.165e−02
	$9p\,^2P^0_{3/2} - 10d\,^2D_{5/2}$	1.886e−02	$9p\,^2P^0_{3/2} - 11d\,^2D_{5/2}$	6.940e−03
	$9p\,^2P^0_{3/2} - 12d\,^2D_{5/2}$	1.070e−03	$10p\,^2P^0_{3/2} - 9d\,^2D_{5/2}$	1.336e−07
	$10p\,^2P^0_{3/2} - 10d\,^2D_{5/2}$	5.970e−03	$10p\,^2P^0_{3/2} - 11d\,^2D_{5/2}$	1.173e−02
	$10p\,^2P^0_{3/2} - 12d\,^2D_{5/2}$	5.615e−03	$5d\,^2D_{5/2} - 4f\,^2F^0_{7/2}$	1.412e−01
	$5d\,^2D_{5/2} - 5f\,^2F^0_{7/2}$	7.511e−02	$5d\,^2D_{5/2} - 6f\,^2F^0_{7/2}$	5.661e−03
	$6d\,^2D_{5/2} - 5f\,^2F^0_{7/2}$	5.354e−02	$6d\,^2D_{5/2} - 6f\,^2F^0_{7/2}$	5.084e−02
	$7d\,^2D_{5/2} - 6f\,^2F^0_{7/2}$	2.327e−02	$5s\,^2S_{1/2} - 6f\,^2P^0_{1/2}$	1.881e−10
	$5s\,^2S_{1/2} - 5p\,^2P^0_{1/2}$	1.558e+00	$5s\,^2S_{1/2} - 6p\,^2P^0_{1/2}$	9.082e−01
	$5s\,^2S_{1/2} - 7p\,^2P^0_{1/2}$	2.088e−02	$5s\,^2S_{1/2} - 8p\,^2P^0_{1/2}$	1.179e−04
	$5s\,^2S_{1/2} - 9p\,^2P^0_{1/2}$	2.618e−07	$5s\,^2S_{1/2} - 10p\,^2P^0_{1/2}$	2.881e−10
	$6s\,^2S_{1/2} - 6p\,^2P^0_{1/2}$	1.136e−01	$5s\,^2S_{1/2} - 7p\,^2P^0_{1/2}$	2.799e−01
	$6s\,^2S_{1/2} - 8p\,^2P^0_{1/2}$	3.501e−02	$5s\,^2S_{1/2} - 9p\,^2P^0_{1/2}$	1.120e−03

(continued)

Table 4.40 (continued)

Atom	Transition	WBEPM theory	Transition	WBEPM theory
	$6s\,^2S_{1/2} - 10p\,^2P^0_{1/2}$	1.440e−05	$7s\,^2S_{1/2} - 7p\,^2P^0_{1/2}$	2.343e−02
	$7s\,^2S_{1/2} - 8p\,^2P^0_{1/2}$	1.027e−01	$7s\,^2S_{1/2} - 9p\,^2P^0_{1/2}$	2.794e−02
	$7s\,^2S_{1/2} - 10p\,^2P^0_{1/2}$	2.126e−03	$8s\,^2S_{1/2} - 8p\,^2P^0_{1/2}$	7.186e−03
	$8s\,^2S_{1/2} - 9p\,^2P^0_{1/2}$	4.456e−02	$8s\,^2S_{1/2} - 10p\,^2P^0_{1/2}$	1.973e−02
	$9s\,^2S_{1/2} - 9p\,^2P^0_{1/2}$	2.778e−03	$9s\,^2S_{1/2} - 10p\,^2P^0_{1/2}$	2.181e−02
	$10s\,^2S_{1/2} - 10p\,^2P^0_{1/2}$	1.250e−03		
Au I	$6s\,^2S_{1/2} - 6p\,^2P^0_{3/2}$	2.842e+00	$6s\,^2S_{1/2} - 7p\,^2P^0_{3/2}$	9.798e−01
	$6s\,^2S_{1/2} - 8p\,^2P^0_{3/2}$	1.787e−02	$7s\,^2S_{1/2} - 7p\,^2P^0_{3/2}$	1.74e−01
	$7s\,^2S_{1/2} - 8p\,^2P^0_{3/2}$	3.184e−01	$8s\,^2S_{1/2} - 8p\,^2P^0_{3/2}$	2.185e−02
	$6p\,^2P^0_{3/2} - 6d\,^2D_{5/2}$	8.123e−01	$6p\,^2P^0_{3/2} - 7d\,^2D_{5/2}$	9.95e−02
	$6p\,^2P^0_{3/2} - 8d\,^2D_{5/2}$	1.692e−03	$6p\,^2P^0_{3/2} - 9d\,^2D_{5/2}$	9.92e−06
	$6p\,^2P^0_{3/2} - 10d\,^2D_{5/2}$	3.003e−08	$6p\,^2P^0_{3/2} - 11d\,^2D_{5/2}$	4.897e−11
	$6p\,^2P^0_{3/2} - 12d\,^2D_{5/2}$	4.828e−14	$6p\,^2P^0_{3/2} - 13d\,^2D_{5/2}$	3.095e−17
	$6p\,^2P^0_{3/2} - 14d\,^2D_{5/2}$	1.424e−20	$7p\,^2P^0_{3/2} - 6d\,^2D_{5/2}$	2.030e−03
	$7p\,^2P^0_{3/2} - 7d\,^2D_{5/2}$	2.78e−01	$7p\,^2P^0_{3/2} - 8d\,^2D_{5/2}$	7.922e−02
	$7p\,^2P^0_{3/2} - 9d\,^2D_{5/2}$	4.788e−03	$7p\,^2P^0_{3/2} - 10d\,^2D_{5/2}$	1.195e−04
	$7p\,^2P^0_{3/2} - 11d\,^2D_{5/2}$	1.345e−06	$7p\,^2P^0_{3/2} - 12d\,^2D_{5/2}$	8.343e−09
	$7p\,^2P^0_{3/2} - 13d\,^2D_{5/2}$	3.147e−11	$7p\,^2P^0_{3/2} - 14d\,^2D_{5/2}$	8.026e−14
	$8p\,^2P^0_{3/2} - 7d\,^2D_{5/2}$	7.663e−04	$8p\,^2P^0_{3/2} - 8d\,^2D_{5/2}$	8.052e−02
	$8p\,^2P^0_{3/2} - 9d\,^2D_{5/2}$	4.759e−02	$8p\,^2P^0_{3/2} - 10d\,^2D_{5/2}$	6.440e−03
	$8p\,^2P^0_{3/2} - 11d\,^2D_{5/2}$	3.296e−04	$8p\,^2P^0_{3/2} - 12d\,^2D_{5/2}$	8.81e−06
	$8p\,^2P^0_{3/2} - 13d\,^2D_{5/2}$	1.174e−07	$8p\,^2P^0_{3/2} - 14d\,^2D_{5/2}$	1.060e−09
	$6d\,^2D_{5/2} - 5f\,^2F^0_{7/2}$	1.414e−01	$6s\,^2S_{1/2} - 6p\,^2P^0_{1/2}$	2.162e+00
	$6s\,^2S_{1/2} - 7p\,^2P^0_{1/2}$	1.201e+00	$6s\,^2S_{1/2} - 8p\,^2P^0_{1/2}$	2.567e−02
	$7s\,^2S_{1/2} - 7p\,^2P^0_{1/2}$	8.949e−02	$7s\,^2S_{1/2} - 8p\,^2P^0_{1/2}$	3.431e−01
	$8s\,^2S_{1/2} - 8p\,^2P^0_{1/2}$	1.387e−02		

Data source Zheng NW, Wang T, Yang RY. J Chem Phys, 2000, 113:6169

Notes [1] In the table the results of Cu I are obtained without adjusting the number of nodes, while the results of Ag I and Au I are obtained with the number of nodes adjusted. For Ag I and Au I, it is necessary to adjust the number of nodes; the similar situations are also shown in the calculation of transition probability of alkali metal atoms. Li atom needs no adjustment, while for Na, K, Ru and Cs, it is necessary to adjust the number of nodes. There are no consistent views about in what situation the number of nodes needs to adjust or doesn't need to adjust, how to adjust and what the mechanism of the adjustment is. Readers can refer to the following literature if interested: (a) Kahn JR, Baybutt P, Truhlar DG. J Chem Phys, 1976, 65:3826. (b) Kostelecky VA, Nieto MM. Phys Rev A, 1985, 32:1293; Phys Rev Lett, 1984, 53:2285

4.3 Calculation of Oscillator Strength, Transition Probability ...

Table 4.41 Transition probability (10^8 s^{-1}) for C I–C IV

Atom or ion	Transition	$T_{WBEPM, calc}$	Acceptable value[①]
C I [He]$2s^2 2p$ ($^2P^0$) nl	$3s\,^1P_1^0 - 3p\,^1D_2$	0.275 940	0.291
	$3s\,^1P_1^0 - 4p\,^1D_2$	0.023 881	0.026 0
	$5s\,^1P_1^0 - 3p\,^1D_2$	0.042 477	0.044 3
	$5s\,^3P_1^0 - 3p\,^3D_2$	0.031 737	0.031 2
	$5s\,^3P_2^0 - 3p\,^3D_3$	0.035 135	0.032 6
	$6s\,^3P_0^0 - 3p\,^3D_1$	0.021 724	0.021 3
	$6s\,^3P_1^0 - 3p\,^3D_1$	0.005 408	0.005 34
	$6s\,^3P_1^0 - 3p\,^3D_2$	0.016 213	0.016 0
	$6s\,^3P_2^0 - 3p\,^3D_2$	0.003 174	0.003 22
	$6s\,^3P_2^0 - 3p\,^3D_3$	0.017 765	0.017 9
	$7s\,^3P_1^0 - 3p\,^3D_1$	0.003 119	0.003 04
	$3p\,^3D_1 - 4d\,^3F_2^0$	0.023 088	0.021 7
	$3p\,^3D_2 - 4d\,^3F_3^0$	0.025 155	0.021 9
	$3p\,^3D_3 - 4d\,^3F_4^0$	0.031 008	0.024 7
C II [He]$2s^2$ (^1S) nl	$3s\,^2S_{1/2} - 3p\,^2P_{1/2}^0$	0.428 271	0.362
	$3s\,^2S_{1/2} - 3p\,^2P_{3/2}^0$	0.429 240	0.363
	$3s\,^2S_{1/2} - 4p\,^2P_{1/2}^0$	0.204 401	0.231
	$3s\,^2S_{1/2} - 4p\,^2P_{3/2}^0$	0.203 244	0.231

(continued)

Table 4.41 (continued)

Atom or ion	Transition	$T_{\text{WBEPM, calc}}$	Acceptable value[1]
	$3p\,^2P^0_{1/2}$ – $3d\,^2D_{3/2}$	0.380 269	0.352
	$3p\,^2P^0_{3/2}$ – $3d\,^2D_{3/2}$	0.075 885	0.070 3
	$3p\,^2P^0_{3/2}$ – $3d\,^2D_{5/2}$	0.455 466	0.422
C III [He]2s (^{2}S) nl	$2p\,^1P^0_1$ – $3d\,^1D_2$	62.673 71	62.4
	$3p\,^1P^0_1$ – $3d\,^1D_2$	0.499 261	0.427
	$3s\,^3S_1$ – $3p\,^3P^0_0$	0.782 970	0.724
	$3s\,^3S_1$ – $3p\,^3P^0_1$	0.783 609	0.725
	$3s\,^3S_1$ – $3p\,^3P^0_2$	0.785 084	0.726
	$2p\,^3P^0_1$ – $3d\,^3D_2$	54.316 79	59.1
	$2p\,^3P^0_2$ – $3d\,^3D_3$	72.416 62	79.7
	$3p\,^3P^0_0$ – $3d\,^3D_1$	0.050 024	0.044 0
	$3p\,^3P^0_1$ – $3d\,^3D_1$	0.037 456	0.032 9
	$3p\,^3P^0_1$ – $3d\,^3D_2$	0.067 444	0.059 3
	$3p\,^3P^0_2$ – $3d\,^3D_1$	0.002 488	0.002 19
	$3p\,^3P^0_2$ – $3d\,^3D_2$	0.022 396	0.019 7
	$3p\,^3P^0_2$ – $3d\,^3D_3$	0.089 662	0.078 8
C IV 1s^2 (^{1}S) nl	$2s\,^2S_{1/2}$ – $2p\,^2P^0_{1/2}$	2.649 110	2.64
	$2s\,^2S_{1/2}$ – $2p\,^2P^0_{3/2}$	2.663 319	2.65

(continued)

4.3 Calculation of Oscillator Strength, Transition Probability …

Table 4.41 (continued)

Atom or ion	Transition	$T_{\text{WBEPM, calc}}$	Acceptable value[①]
	$2s\,^2S_{1/2}$ – $3p\,^2P^0_{1/2}$	43.521 02	46.3
	$2s\,^2S_{1/2}$ – $3p\,^2P^0_{3/2}$	43.453 96	46.3
	$3s\,^2S_{1/2}$ – $3p\,^2P^0_{1/2}$	0.314 645	0.316
	$3s\,^2S_{1/2}$ – $3p\,^2P^0_{3/2}$	0.316 437	0.317
	$2p\,^2P^0_{3/2}$ – $3d\,^2D_{5/2}$	176.030 0	176

Data source Zheng NW, Wang T, Ma DX, et al. J Opt Soc Am B, 2001, 18:1395
Notes [①] Acceptable values are taken from Lide DR. CRC handbook of chemistry and physics, 81st edn. CRC Press, Inc., Boca Raton, Florida, 1999: pp 10-88–10-146. [Computed files]: CRC net Base

Table 4.42 The comparison of calculated oscillator strength of transitions between excited states with literature data for magnesium-like ions (P IV, S V, Cl VI, Ar VII)

Ion	Transition	Oscillator strength	
		$T_{\text{WBEPM, calc}}$	Literature data
P IV	$3s4s\,^3S_1$ – $3s4p\,^3P^0_0$	0.128 9	0.12[c]
	$3s4s\,^3S_1$ – $3s4p\,^3P^0_1$	0.387 3	0.35[c]
	$3s4s\,^3S_1$ – $3s4p\,^3P^0_2$	0.648 6	0.66[c]
	$3s3p\,^3P^0_0$ – $3s3d\,^3D_1$	0.759 3	
	$3s3p\,^3P^0_0$ – $3s4d\,^3D_1$	0.004 128	
	$3s3p\,^3P^0_0$ – $3s4d\,^3D_1$	0.189 7	
	$3s3p\,^3P^0_1$ – $3s3d\,^3D_2$	0.569 2	
	$3s3p\,^3P^0_1$ – $3s4d\,^3D_1$	0.001 006	
	$3s3p\,^3P^0_1$ – $3s4d\,^3D_2$	0.003 025	
	$3s3p\,^3P^0_2$ – $3s3d\,^3D_1$	0.007 582	

(continued)

Table 4.42 (continued)

Ion	Transition	Oscillator strength	
		$T_{\text{WBEPM, calc}}$	Literature data
	$3s3p\ ^3P_2^0 - 3s3d\ ^3D_2$	0.113 7	
	$3s3p\ ^3P_2^0 - 3s3d\ ^3D_3$	0.636 9	
	$3s3p\ ^3P_2^0 - 3s4d\ ^3D_1$	3.808E−05	
	$3s3p\ ^3P_2^0 - 3s4d\ ^3D_2$	0.000 573 0	
	$3s3p\ ^3P_2^0 - 3s4d\ ^3D_3$	0.003 222	
	$3s3d\ ^3D_1 - 3s4p\ ^3P_0^0$	0.075 42	0.075[c]
	$3s3d\ ^3D_1 - 3s4p\ ^3P_1^0$	0.056 49	0.056[c]
	$3s3d\ ^3D_1 - 3s4p\ ^3P_2^0$	0.003 753	0.003 8[c]
	$3s3d\ ^3D_2 - 3s4p\ ^3P_1^0$	0.101 7	0.10[c]
	$3s3d\ ^3D_2 - 3s4p\ ^3P_2^0$	0.033 78	0.034[c]
	$3s3d\ ^3D_3 - 3s4p\ ^3P_2^0$	0.135 1	0.14[c]
	$3s4p\ ^3P_0^0 - 3s4d\ ^3D_1$	1.146	
	$3s4p\ ^3P_1^0 - 3s4d\ ^3D_1$	0.286 3	
	$3s4p\ ^3P_1^0 - 3s4d\ ^3D_2$	0.858 9	
	$3s4p\ ^3P_2^0 - 3s4d\ ^3D_1$	0.011 44	
	$3s4p\ ^3P_2^0 - 3s4d\ ^3D_2$	0.171 6	
	$3s4p\ ^3P_2^0 - 3s4d\ ^3D_3$	0.960 9	
S V	$3s4s\ ^3S_1 - 3s4p\ ^3P_0^0$	0.119 8	0.103[c], 0.101 5[f]
	$3s4s\ ^3S_1 - 3s4p\ ^3P_1^0$	0.359 8	0.284[c], 0.259 2[f]

(continued)

Table 4.42 (continued)

Ion	Transition	Oscillator strength	
		$T_{\text{WBEPM, calc}}$	Literature data
	$3s4s\,^3S_1 - 3s4p\,^3P_2^0$	0.604 6	0.52[c], 0.530 3[f]
	$3s3p\,^3P_0^0 - 3s3d\,^3D_1$	0.706 8	0.72[g]
	$3s3p\,^3P_1^0 - 3s3d\,^3D_1$	0.176 5	0.18[g]
	$3s3p\,^3P_1^0 - 3s3d\,^3D_2$	0.529 5	0.537[g]
	$3s3p\,^3P_2^0 - 3s3d\,^3D_1$	0.007 044	0.007 2[g]
	$3s3p\,^3P_2^0 - 3s3d\,^3D_2$	0.105 7	0.108[g]
	$3s3p\,^3P_2^0 - 3s3d\,^3D_3$	0.591 7	0.60[g]
	$3s3d\,^3D_1 - 3s4p\,^3P_0^0$	0.063 41	0.032[c], 0.028 7[f]
	$3s3d\,^3D_1 - 3s4p\,^3P_1^0$	0.047 52	0.03[c], 0.023 83[f]
	$3s3d\,^3D_1 - 3s4p\,^3P_2^0$	0.003 146	0.008[c], 0.003 66[f]
	$3s3d\,^3D_2 - 3s4p\,^3P_1^0$	0.085 54	0.058[c], 0.034 36[f]
	$3s3d\,^3D_2 - 3s4p\,^3P_2^0$	0.028 32	0.046[c], 0.025 92[f]
	$3s3d\,^3D_3 - 3s4p\,^3P_2^0$	0.113 3	0.06[c], 0.069 0[f]
	$3s4p\,^3P_0^0 - 3s4d\,^3D_1$	1.060	
	$3s4p\,^3P_1^0 - 3s4d\,^3D_1$	0.265 0	
	$3s4p\,^3P_1^0 - 3s4d\,^3D_2$	0.795 1	
	$3s4p\,^3P_2^0 - 3s4d\,^3D_1$	0.010 56	
	$3s4p\,^3P_2^0 - 3s4d\,^3D_2$	0.158 5	
	$3s4p\,^3P_2^0 - 3s4d\,^3D_3$	0.887 8	

(continued)

Table 4.42 (continued)

Ion	Transition	Oscillator strength	
		$T_{\text{WBEPM, calc}}$	Literature data
Cl VI	$3s4s\,^3S_1 - 3s4p\,^3P^0_0$	0.111 7	0.100^c, $0.099\,2^f$, $0.100\,7^h$
	$3s4s\,^3S_1 - 3s4p\,^3P^0_1$	0.336 2	0.312^c, $0.298\,4^f$, $0.302\,8^h$
	$3s4s\,^3S_1 - 3s4p\,^3P^0_2$	0.565 4	0.507^c, 0.502^f, $0.510\,3^h$
	$3s3p\,^3P^0_0 - 3s3d\,^3D_1$	0.643 6	0.64^g, $0.650\,7^h$
	$3s3p\,^3P^0_1 - 3s3d\,^3D_1$	0.160 6	0.16^g, $0.162\,3^h$
	$3s3p\,^3P^0_1 - 3s3d\,^3D_2$	0.481 8	0.477^g, $0.487\,1^h$
	$3s3p\,^3P^0_2 - 3s3d\,^3D_1$	0.006 399	$0.006\,4^g$, $0.006\,458^h$
	$3s3p\,^3P^0_2 - 3s3d\,^3D_2$	0.095 99	0.094^g, $0.096\,91^h$
	$3s3p\,^3P^0_2 - 3s3d\,^3D_3$	0.537 7	0.53^g, $0.542\,6^i$
	$3s3d\,^3D_1 - 3s4p\,^3P^0_0$	0.054 06	0.053^c, $0.051\,47^f$, $0.051\,35^i$
	$3s3d\,^3D_1 - 3s4p\,^3P^0_1$	0.040 41	0.040^c, $0.038\,3^f$, $0.038\,25^i$
	$3s3d\,^3D_1 - 3s4p\,^3P^0_2$	0.002 673	0.003^c, $0.002\,53^f$, $0.002\,542^i$
	$3s3d\,^3D_2 - 3s4p\,^3P^0_1$	0.072 76	0.072^c, $0.069\,38^f$, $0.069\,24^i$
	$3s3d\,^3D_2 - 3s4p\,^3P^0_2$	0.024 06	0.024^c, $0.022\,9^f$
	$3s3d\,^3D_3 - 3s4p\,^3P^0_2$	0.096 29	0.095^c, $0.091\,96^f$, $0.092\,09^i$
	$3s4p\,^3P^0_0 - 3s4d\,^3D_1$	0.953 4	0.816^i
	$3s4p\,^3P^0_1 - 3s4d\,^3D_1$	0.237 9	$0.202\,8^i$
	$3s4p\,^3P^0_1 - 3s4d\,^3D_2$	0.714 0	$0.610\,2^i$

(continued)

Table 4.42 (continued)

Ion	Transition	Oscillator strength	
		$T_{\text{WBEPM, calc}}$	Literature data
	$3s4p\,^3P_2^0 - 3s4d\,^3D_1$	0.009 471	0.008 06[i]
	$3s4p\,^3P_2^0 - 3s4d\,^3D_2$	0.142 1	0.121 2[i]
	$3s4p\,^3P_2^0 - 3s4d\,^3D_3$	0.796 7	0.681 0[i]
Ar VII	$3s3p\,^3P_0^0 - 3s3d\,^3D_1$	0.592 3	0.57[g], 0.610[i]
	$3s3p\,^3P_1^0 - 3s3d\,^3D_1$	0.147 7	0.14[g], 0.152[j]
	$3s3p\,^3P_1^0 - 3s3d\,^3D_2$	0.443 1	0.43[g], 0.456[j]
	$3s3p\,^3P_2^0 - 3s3d\,^3D_1$	0.005 874	0.005 6[g], 0.006 1[j]
	$3s3p\,^3P_2^0 - 3s3d\,^3D_2$	0.088 13	0.084[g], 0.091[j]
	$3s3p\,^3P_2^0 - 13s3d\,^3D_3$	0.493 7	0.474[g], 0.509[j]

Data source Fan J, Zheng NW. Chem Phys Lett, 2004, 400:273
Notes c. Aashamar K, Luke TM, Talman JD. Phys Scr, 1988, 37:13
f. Godefroid M, Fischer CF. Phys Scr, 1985, 31:237
g. Fawcett BC. At Data Nucl Data, 1983, 28:579
h. Butler K, Mendoza C, Zeippen CJ. J Phys B, 1993, 26:4409
i. Neerja, Gupa GP, Tripathi AN, Msezane AZ. At Data Nucl Data, 2003, 84:85
j. Tayal SS. J Phys B, 1986, 19:3421

4.4 Calculation of Total Electron Encrgy [1, 159, 160]

In the theoretical treatment of many-electron atomic and molecular systems, due to the existence of r_{ij}^{-1}, variables can't be separated and Schrodinger equation can't be solved exactly. Thus, for many-electron systems which have electron–electron interactions, it is common to treat them approximately. The meaning of approximate treatment includes modeling the real physical problems, making it simple and easy to solve; calculating interesting problems or simplified models with approximation.

The key points of the weakest bound electron theory that was stated in Chap. 2 have already shown that the process of removing electrons and the process of adding electrons are mutually reversible and removing electrons and adding electrons reveal two modes for dealing with problems of N-electron atoms or molecules. The mode of adding electrons means that electrons can be handled as a whole in a system, i.e.

Table 4.43 The comparison of calculated oscillator strength of transitions between spectral terms with literature data for C atom

Transition	Calculated value of WBEPM theory	Literature data
$2p^2\ {}^3P - 2p3s\ {}^3P^0$	0.153 2	0.147 4[a]; 0.139 4[b]; 0.160 7[c]; 0.050 4[d]; 0.090 0[e]; 0.170[f]; 0.140[g]; 0.13[h]
$2p^2\ {}^3P - 2p4s\ {}^3P^0$	0.022 64	0.021 71[a]; 0.021 80[b]; 0.015 9[c]; 0.008 86[d]; 0.015 2[e]; 0.020 0[f]; 0.018 9[g]; 0.027[h]
$2p^2\ {}^3P - 2p3d\ {}^3D^0$	0.082 50	0.109 8[a]; 0.094 39[b]; 0.141 4[c]; 0.074 3[d]; 0.075[e]; 0.063 0[f]; 0.094 0[g]; 0.10[h]
$2p^2\ {}^1D - 2p3s\ {}^1P^0$	0.112 0	0.120 8[a]; 0.113 8[b]; 0.316 9[c]; 0.057 3[d]; 0.101[e]; 0.082 0[f]; 0.114[g]; 0.13[h]
$2p^2\ {}^1D - 2p3d\ {}^1F^0$	0.082 68	0.096 08[a]; 0.080 70[b]; 0.315 6[c]; 0.081 8[d]; 0.085[e]; 0.093 0[f]; 0.085 0[g]; 0.094[h]
$2p^2\ {}^1S - 2p3s\ {}^1P^0$	0.099 20	0.085 45[a]; 0.086 20[b]; 0.374 1[c]; 0.062 1[d]; 0.113[e]; 0.094 0[f]
$2p^2\ {}^1S - 2p3d\ {}^1P^0$	0.088 53	0.140 3[a]; 0.132 2[b]; 0.482 3[c]; 0.116[d]; 0.120[e]; 0.120[f]; 0.110[g]; 0.35[h]
$2p3s\ {}^3P^0 - 2p3p\ {}^3S$	0.112 9	0.107 1[a]; 0.107 0[b]; 0.117 2[c]; 0.136[d]; 0.108 5[e]; 0.100[f]
$2p3s\ {}^3P^0 - 2p3p\ {}^3P$	0.351 5	0.359 6[a]; 0.356 4[b]; 0.365 6[c]; 0.434[d]; 0.384 4[e]; 0.310[f]
$2p3s\ {}^3P^0 - 2p3p\ {}^3D$	0.517 1	0.497 1[a]; 0.507 3[b]; 0.526 8[c]; 0.615[d]; 0.538[e]; 0.500[f]
$2p3s\ {}^1P^0 - 2p3p\ {}^1S$	0.122 3	0.121 3[a]; 0.117 4[b]; 0.126 1[c]; 0.158[d]; 0.124 1[e]; 0.110[f]
$2p3s\ {}^1P^0 - 2p3p\ {}^1P$	0.249 2	0.252 0[a]; 0.267 0[b]; 0.246 1[c]; 0.272[d]; 0.302 7[e]
$2p3s\ {}^3P^0 - 2p3p\ {}^1D$	0.610 2	0.636 7[a]; 0.624 0[b]; 0.603 3[c]; 0.701[d]; 0.641[e]; 0.420[f]
$2p3p\ {}^3S - 2p3d\ {}^3P^0$	0.964 8	0.621 7[a]; 0.563 7[b]; 1.017 0[c]; 0.997[d]; 0.860[e]; 0.960[f]
$2p3p\ {}^3P - 2p3d\ {}^3P^0$	0.244 7	0.301 7[a]; 0.303 8[b]; 0.254 9[c]; 0.226[d]; 0.228 1[e]; 0.249[f]
$2p3p\ {}^3P - 2p3d\ {}^3D^0$	0.663 4	0.703 5[a]; 0.693 7[b]; 0.675 1[c]; 0.593[d]; 0.678[e]
$2p3p\ {}^3D - 2p3d\ {}^3D^0$	0.134 2	0.148 7[a]; 0.142 6[b]; 0.140 2[c]; 0.149[d]; 0.147 1[e]; 0.132[f]
$2p3p\ {}^3D - 2p3d\ {}^3F^0$	0.746 5	0.812 7[a]; 0.783 1[b]; 0.776 7[c]; 0.830[d]; 0.814[e]; 0.700[f]

(continued)

4.4 Calculation of Total Electron Energy

Table 4.43 (continued)

Transition	Calculated value of WBEPM theory	Literature data
2p3p ^{1}P – 2p3d ^{1}D^0	0.635 6	0.696 0[a]; 0.670 6[b]; 0.663 9[c]; 0.770[d]; 0 714[e]
2p3p ^{1}D – 2p3d ^{1}D^0	0.108 9	0.115 9[a]; 0.114 4[b]; 0.111 8[c]; 0.087 2[d]; 0.112 6[e]
2p3p ^{1}D – 2p3d ^{1}F^0	0.731 7	0.785 0[a]; 0.769 5[b]; 0.735 1[c]; 0.566[d]; 0.761[e]; 0.740[f]

Data source Zheng NW, Wang T. Astrophys J Suppl Ser, 2002, 143:231
Notes
a. Luo D, Pradhan AKJ. J Phys B, 1989, 22:3377
b. Nussbaumer H, Storey PJ. A & A, 1984, 140:383
c. Victor GA, Escalante V. At Data Nucl Data Tables, 1988, 40:203
d. Hofsaess D. J Quant Spectrosc Radiat Transfer, 1982, 28:131
e. McEachran RP, Cohen M. J Quant Spectrosc Radiat Transfer, 1982, 27:119
f. Wiese WL, Smith MW, Glennon M. Atomic transition probabilities (NSRDS-NBS4: Washington: USGPO), 1966
g. Goldbach C, Martin M, Nollez G. A & A, 1989, 221:155; Goldbach C, Nollez G. A & A, 1987, 181:203
h. Haar RR, et al. A & A, 1991, 241:321

commonly called *N*-electron problem. The existing quantum theoretical methods, such as self-consistent field method, molecular orbital theory, etc., have achieved great success and abundant accomplishments. The mode of adding electrons opens a channel for WBE theory to take in and integrate with these accomplishments and methods, while the mode of removing electrons finds theoretical foundations for separability of electrons, manifesting the character of electrons, and also opens a channel for WBE theory to take in and communicate with the existing quantum theories and accomplishments which are based on separability and to look for some rules in chemistry and physics.

For atomic systems, three different approximate methods will be introduced in the following: ① calculation of total electron energy using the ionization energy of the weakest bound electron; ② variational treatment within WBEPM Theory; ③ perturbation treatment within WBEPM Theory.

4.4.1 Calculation of Total Electron Energy of the System Using Ionization Energy

In Sect. 4.1, the following formula has been given

$$E_{el} = -\sum_{\mu=1}^{N} I_\mu \qquad (4.4.1)$$

This formula indicates that the total electron energy of a ground-state N-electron system is equal to the negative of the sum of successive ionization energies of the N-electron system.

$$I_{cal} = \frac{R}{n'^2}[(Z-\sigma)^2 + g(Z-Z_0)] + \sum_{i=0}^{4} a_i Z^i \quad (4.4.2)$$

This is the formula for calculating ionization energy (ground state and excited state) for an iso-spectrum-level series of an atomic system using WBEPM Theory after considering correlation and relativistic correction to some extent.

Substituting the successive ionization energies calculated by Eq. (4.4.2) into Eq. (4.4.1), the total electron energy of the atomic system can be obtained. For instance, the calculated total electron energies for element $Z = 2 - 8$ are listed in Tables 4.44 and 4.45.

4.4.2 Variational Treatment on the Energy of the He-Sequence Ground State with the WBEP Theory

Helium atom is a two-electron system and it is one of the simplest and the most representative many-electron systems. Treating helium atomic system plays an important role in the development of quantum mechanics theories, and now variation treatment and perturbation treatment of ground-state helium have become examples in quantum mechanics and quantum chemistry works. And so far, the interests in the high-precision quantum mechanics treatment of some small representative many-electron atomic systems such as helium atom and lithium atom are still maintained [161–165]. Through this type of study on basic theory and methodology, one may be able to deeply understand the role of all kinds of interactions of many-electron systems in the theory of quantum electronic structures, and their influence on the properties of the system and possible practical applications.

Here, we will also use WBEPM Theory for the variational treatment of helium sequence, i.e. helium atoms and helium-like ions to build a general method for variational treatment of many-electron atomic systems using WBEPM Theory, and at the same time to expect to obtain some good results.

4.4.2.1 Single Generalized Laguerre Function as a Trial Function

All the members of He^+-sequence are hydrogen-like ions with only one 1s electron. This electron is the weakest bound electron of the present system. Since a hydrogen-like system can be solved exactly, the spatial function of its spin orbital is hydrogen-like wave function.

4.4 Calculation of Total Electron Energy

Table 4.44 The total electron energy E_{el} of an atomic system calculated from successive ionization energy $I_{\mu,cal}$ (eV)[①]

Z	Element	Successive ionization energy								E_{el}
		I_1	I_2	I_3	I_4	I_5	I_6	I_7	I_8	
2	He	24.654 80	54.424[a]							79.078 8
3	Li	5.305 74	75.635 57	122.454[a]						203.395 31
4	Be	9.232 22	18.235 22	153.857 60	217.696[a]					399.021
5	B	8.229 85	25.172 54	37.980 61	259.325 02	340.15[a]				670.858
6	C	11.160 62	24.435 12	47.938 82	64.544 37	392.046 91	489.816[a]			1 029.942
7	N	14.280 80	29.632 92	47.504 63	77.531 40	97.930 58	552.037 35	666.694[a]		1 485.612
8	O	13.406 87	35.182 43	54.983 41	77.438 63	113.952 33	138.144 95	739.315 43	870.784[a]	2 043.208

Notes [①] The values of a are calculated using the energy formula for hydrogen-like atoms; the rest of ionization energies in the table is taken from the literature: Zheng NW, Zhou T, Wang T, et al. Phys Rev A, 2002, 65:052510

Table 4.45 The comparison of calculated E_{el} using WBEPM theory with other theoretical methods and experimental data (eV)

Z	Element	E_{el}		
		WBEPM calc[a]	HF value[b]	Exp. value[c]
2	He	79.078 8	77.871	79.005 147
3	Li	203.395 31	202.256 42	203.486 009
4	Be	399.021	396.555 33	399.149 1
5	B	670.858	667.475 1	670.984 47
6	C	1 029.942	1 025.567 7	1 030.105 64
7	N	1 485.612	1 480.336 6	1 486.066 14
8	O	2 043.208	2 035.683 3	2 043.806 98

Notes
a. The values are taken from Table 4.44
b. The values are taken from Roetti CE. Chem Phys, 1974, 60:4725. The unit of HF values in this literature is a.u., and they can be converted to eV by using 1 a.u. = 27.211 608 eV
c. The values are taken from Lide DR. CRC handbook of chemistry and physics: 2005–2006, 86th edn. Taylor & Francis, New York, 2006: 10-202–204

$$\psi_{1s}(r) = \frac{1}{\sqrt{\pi}} Z^{3/2} e^{-Zr} \quad (4.4.3)$$

When adding one electron into the system of He$^+$-sequence to make it the He-sequence ground state, the inner system will be reorganized due to the interactions of the newly added electron with old electrons [166, 167]. For one-electron wave functions (or one-electron states) of two electrons of He-sequence after reorganization, neither of them is identical to the wave function (or one-electron state) of that electron in the original He$^+$-sequence, (if there is one that is identical, that is the image given by frozen orbital in Koopman's theorem). Therefore, we use single generalized Laguerre function given in Chap. 3 as a trial function to formulate the spatial part of the electron spin orbital. Here, we are in no hurry to take the same form for the spatial orbitals of two electrons. Thus, the antisymmetry electron wave function for the He-sequence ground state system is,

$$\Psi = (2!)^{-1/2} \sum_p (-1)^p P\{\psi_I(1)\alpha(1)\psi_{II}(2)\beta(2)\} \quad (4.4.4)$$

where

$$\psi_I(1) = R_{n'_1 l'_1} Y_{00}(\theta_1 \phi_1)$$
$$= \frac{1}{\sqrt{4\pi \Gamma(2l'_1 + 3)}} \left(\frac{2Z'_1}{n'_1}\right)^{l'_1 + 3/2} r^{l'_1} e^{-Z'_1 r_1 / n'_1} \quad (4.4.5)$$

4.4 Calculation of Total Electron Energy

$$\psi_{II}(2) = R_{n'_2 l'_2} Y_{00}(\theta_2 \phi_2)$$

$$= \frac{1}{\sqrt{4\pi \Gamma(2l'_2 + 3)}} \left(\frac{2Z'_2}{n'_2}\right)^{l'_1+3/2} r^{l'_1} e^{-Z'_2 r_2 / n'_2} \qquad (4.4.6)$$

The non-relativistic Hamiltonian of the system is

$$\hat{H} = -\frac{1}{2}\nabla_1^2 - \frac{1}{2}\nabla_2^2 - \frac{Z}{r_1} - \frac{Z}{r_2} + \frac{1}{r_{12}} \qquad (4.4.7)$$

Then, the approximate expectation value of the total electron energy of the system is

$$W = \frac{\langle \Psi | \hat{H} | \Psi \rangle}{\langle \Psi | \Psi \rangle} \qquad (4.4.8)$$

Substituting Eq. (4.4.4) into Eq. (4.4.6), then we get

$$\langle \Psi | \Psi \rangle = 1 \qquad (4.4.9)$$

In Eq. (4.4.8)

$$\langle \Psi | \hat{H} | \Psi \rangle$$

$$= \left\langle \Psi \left| -\frac{1}{2}\nabla_1^2 - \frac{1}{2}\nabla_2^2 - \frac{Z}{r_1} - \frac{Z}{r_2} + \frac{1}{r_{12}} \right| \Psi \right\rangle$$

$$= \left\langle \Psi \left| -\frac{1}{2}\nabla_1^2 \right| \Psi \right\rangle + \left\langle \Psi \left| -\frac{1}{2}\nabla_2^2 \right| \Psi \right\rangle + \left\langle \Psi \left| -\frac{Z}{r_1} \right| \Psi \right\rangle$$

$$+ \left\langle \Psi \left| -\frac{Z}{r_2} \right| \Psi \right\rangle + \left\langle \Psi \left| \frac{1}{r_{12}} \right| \Psi \right\rangle \qquad (4.4.10)$$

The first and second terms on the right side of the above equation are integral of kinetic energy, the third and fourth terms are integral of nuclear attractive energy, and the fifth term is integral of electron repulsive energy. Separate descriptions about calculation of these integrals are as follows:

(1) Integral of kinetic energy

Two calculation methods are introduced.
Method 1: the formula for integral of kinetic energy was given in Ref. [11]

$$T(nl) = \left\langle nlm \left| -\frac{1}{2}\nabla_1^2 \right| nlm \right\rangle$$

$$= \frac{1}{2} \int_0^\infty \left[r^2 \frac{dR_{nl}^*}{dr} \frac{dR_{nl}}{dr} + l(l+1) R_{nl}^* R_{nl} \right] dr \qquad (4.4.11)$$

In the current situation, the integral of kinetic energy $\langle \Psi | -\frac{1}{2}\nabla_1^2 | \Psi \rangle$ and $\langle \Psi | -\frac{1}{2}\nabla_2^2 | \Psi \rangle$ can be calculated by just substituting R_{nl}^* and R_{nl} with $R_{n_1'l_1'}^*$ and $R_{n_1'l_1'}$ or $R_{n_2'l_2'}^*$ and $R_{n_2'l_2'}$. The results are

$$\left\langle \Psi \left| -\frac{1}{2}\nabla_1^2 \right| \Psi \right\rangle = \frac{Z_1'^2}{n_1'^2}\left(\frac{1}{2} - \frac{l_1'}{2l_1'+1}\right) \tag{4.4.12}$$

and

$$\left\langle \Psi \left| -\frac{1}{2}\nabla_2^2 \right| \Psi \right\rangle = \frac{Z_2'^2}{n_2'^2}\left(\frac{1}{2} - \frac{l_2'}{2l_2'+1}\right) \tag{4.4.13}$$

Method 2: According to the statement in Chap. 3, if the potential function of the weakest bound electron has the form

$$V(r) = -\frac{Z'}{r} + \frac{d(d+1)+2dl}{2r^2} \tag{4.4.14}$$

Then the one-electron Schrodinger equation of the weakest bound electron is

$$\left(-\frac{1}{2}\nabla^2 - \frac{Z'}{r} + \frac{d(d+1)+2dl}{2r^2}\right)\psi = \varepsilon\psi \tag{4.4.15}$$

So

$$\left\langle \psi \left| -\frac{1}{2}\nabla^2 \right| \psi \right\rangle = \langle \psi | \varepsilon | \psi \rangle + \left\langle \psi \left| \frac{Z'}{r} \right| \psi \right\rangle$$
$$+ \left\langle \psi \left| -\left[\frac{d(d+1)+2dl}{2r^2}\right] \right| \psi \right\rangle \tag{4.4.16}$$

It is very easy to get the results for all terms on the right side of Eq. (4.4.16) using the formula of energy for the weakest bound electron which is given in Chap. 3 and the formula of $\langle n'l' | r^k | n'l' \rangle$:

$$\langle \psi | \varepsilon | \psi \rangle = \varepsilon = -\frac{Z'^2}{2n'^2} \tag{4.4.17}$$

$$\left\langle \psi \left| \frac{Z'}{r} \right| \psi \right\rangle = Z'\left\langle \psi \left| \frac{1}{r} \right| \psi \right\rangle = Z'\frac{Z'}{n'^2} = \frac{Z'^2}{n'^2} \tag{4.4.18}$$

$$\left\langle \psi \left| -\left[\frac{d(d+1)+2dl}{2r^2}\right] \right| \psi \right\rangle = -\frac{Z'^2 l'}{n'^2(2l'+1)} \tag{4.4.19}$$

Substituting Eq. (4.4.17) to Eq. (4.4.19) into Eq. (4.4.16), we get

4.4 Calculation of Total Electron Energy

$$\left\langle \psi \left| -\frac{1}{2}\nabla^2 \right| \psi \right\rangle = \frac{Z'^2}{n'^2}\left(\frac{1}{2} - \frac{l'}{2l'+1}\right) \qquad (4.4.20)$$

Further we have

$$\left\langle \Psi \left| -\frac{1}{2}\nabla^2 \right| \Psi \right\rangle = \frac{Z'^2}{n'^2}\left(\frac{1}{2} - \frac{l'}{2l'+1}\right) \qquad (4.4.21)$$

This result and the result by method 1, i.e. Eqs. (4.4.12) and (4.4.13) are completely identical.

(2) Integral of nuclear attractive energy

Using the formula of $\langle n'l'|r^k|n'l'\rangle$ given in Chap. 3, one quickly has

$$\left\langle \Psi \left| -\frac{Z}{r_1} \right| \Psi \right\rangle = -\frac{ZZ'_1}{n_1'^2} \qquad (4.4.22)$$

and

$$\left\langle \Psi \left| -\frac{Z}{r_2} \right| \Psi \right\rangle = -\frac{ZZ'_2}{n_2'^2} \qquad (4.4.23)$$

(3) Integral of electron repulsive energy

Using the expansion of $\frac{1}{r_{12}}$ in the spherical coordinates [86]

$$\frac{1}{r_{12}} = \sum_{l=0}^{\infty}\sum_{m=-l}^{l}\frac{4\pi}{2l+1}\frac{r_<^l}{r_>^{l+1}}Y_{lm}(\theta_1\phi_1)Y_{lm}^*(\theta_2\phi_2) \qquad (4.4.24)$$

Then we have

$$\left\langle \Psi \left| \frac{1}{r_{12}} \right| \Psi \right\rangle = \left\langle \psi_\mathrm{I}(1)\psi_\mathrm{II}(2) \left| \frac{1}{r_{12}} \right| \psi_\mathrm{I}(1)\psi_\mathrm{II}(2) \right\rangle$$

$$= \langle \psi_\mathrm{I}(1)\psi_\mathrm{II}(2)| \sum_{l=0}^{\infty}\sum_{m=-l}^{l}\frac{4\pi}{2l+1}\frac{r_<^l}{r_>^{l+1}}Y_{00}(\theta_1\phi_1)$$
$$\times Y_{00}^*(\theta_2\phi_2)|\psi_\mathrm{I}(1)\psi_\mathrm{II}(2)\rangle \qquad (4.4.25)$$

Substituting Eqs. (4.4.5) and (4.4.6) into it and using orthonormality of spherical harmonic, we get

$$\left\langle \Psi \left| \frac{1}{r_{12}} \right| \Psi \right\rangle$$
$$= \frac{1}{\Gamma(2l'_1+3)\Gamma(2l'_2+3)}\left(\frac{2Z'_1}{n'_1}\right)^{2l'_1+3}\left(\frac{2Z'_2}{n'_2}\right)^{2l'_2+3}$$

$$\times \int_0^\infty \int_0^\infty r_1^{2l_1'} e^{-\frac{2Z_1' r_1}{n_1'}} r_2^{2l_2'} e^{-\frac{2Z_2' r_2}{n_2'}} \frac{1}{r_>} r_1^2 r_2^2 dr_1 dr_2$$

$$= \frac{Z_2'}{n_2'^2} - \frac{\left(\frac{2Z_1'}{n_1'}\right)^{2l_1'+2} \left(\frac{2Z_2'}{n_2'}\right)^{2l_2'+3}}{\Gamma(2l_1'+3)\Gamma(2l_2'+3)} \left(\frac{2Z_1'}{n_1'} + \frac{2Z_2'}{n_2'}\right)^{-(2l_1'+2l_2'+4)}$$

$$\times \Gamma(2l_1'+2l_2'+4) - \frac{\left(\frac{2Z_2'}{n_2'}\right)^{2l_2'+3} \left(\frac{2Z_1'}{n_1'}\right)^{-(2l_2'+2)}}{\Gamma(2l_1'+3)\Gamma(2l_2'+3)}$$

$$\times \int_0^\infty \left[(2l_1'+2)x^{2l_2'+1} - x^{2l_2'+2}\right] \exp\left(-\frac{Z_2' n_1' x}{Z_1' n_2'}\right) \times \Gamma(2l_1'+2, x) dx \quad (4.4.26)$$

Substituting the formulae for integral of kinetic energy, integral of nuclear attractive energy and integral of electron repulsive energy as well as Eq. (4.4.9) into Eqs. (4.4.10) and (4.4.8), we get

$$W = \sum_{i=1}^2 \left[\frac{Z_i'^2}{n_i'^2} \left(\frac{1}{2} - \frac{l_i'}{2l_i'+1} \right) + \frac{-ZZ_i'}{n_i'^2} \right] + \frac{Z_2'}{n_2'^2}$$

$$- \frac{\left(\frac{2Z_1'}{n_1'}\right)^{2l_1'+2} \left(\frac{2Z_2'}{n_2'}\right)^{2l_2'+3}}{\Gamma(2l_1'+3)\Gamma(2l_2'+3)} \left(\frac{2Z_1'}{n_1'} + \frac{2Z_2'}{n_2'}\right)^{-(2l_1'+2l_2'+4)}$$

$$\times \Gamma(2l_1'+2l_2'+4) - \frac{\left(\frac{2Z_2'}{n_2'}\right)^{2l_2'+3} \left(\frac{2Z_1'}{n_1'}\right)^{-(2l_2'+2)}}{\Gamma(2l_1'+3)\Gamma(2l_2'+3)}$$

$$\times \int_0^\infty \left[(2l_1'+2)x^{2l_2'+1} - x^{2l_2'+2}\right] \exp\left(\frac{-Z_2' n_1' x}{Z_1' n_2'}\right)$$

$$\times \Gamma(2l_1'+2, x) dx \quad (4.4.27)$$

$\Gamma(\alpha, x)$ in Eqs. (4.4.26) and (4.4.27) is incomplete gamma function. There are four independent parameters Z_1' and Z_2' as well as d_1' and d_2' ($n_1' = n_1 + d_1, l_1' = l_1 + d_1$; $n_2' = n_2 + d_2, l_2' = l_2 + d_2$) in Eq. (4.4.27), since $\left(\frac{\partial W}{\partial Z_i'}\right)$ and $\left(\frac{\partial W}{\partial d_i}\right)$ can't be written in analytical forms, the minimum of W is searched by numerical analysis. We found that, for He atom, when $Z_1' = Z_2' = 1.53929$ and $d_1 = d_2 = -0.04493$, the minimum of W is equal to $-2.854\ 21$ a.u., and the minimum of W and the corresponding parameters for other members of He-sequence are listed in Table 4.46.

If we let $Z_1' = Z_2' = Z'$, $d_1 = d_2 = d$ (i.e. $n_1' = n_2', l_1' = l_2'$), Eq. (4.4.27) can be written as

$$W = \frac{Z'^2}{n'^2(2l'+1)} - \frac{2ZZ'}{n'^2} + \frac{Z'}{n'^2}\left[1 - 2^{-4n'} \times \frac{\Gamma(4n'+1)}{\Gamma^2(2n'+1)}\right] \quad (4.4.28)$$

4.4 Calculation of Total Electron Energy

Table 4.46 The comparison of the results of WBEPM theory by using single generalized Laguerre polynomial with the results of HF method by using single ζ (Zeta) function (a.u.)

	He I	Li II	Be III	B IV	C V	N VI	O VII	F VIII
Z'_i value ($Z'_1 = Z'_2$)	1.539 29	2.533 4	3.530 61	4.528 98	5.527 91	6.527 15	7.526 59	8.526 16
d_i value ($d_1 = d_2$)	−0.044 93	−0.029 097	−0.021 507	−0.017 055	−0.014 13	−0.012 062	−0.010 521	−0.009 329
Total electron energy (T.E.)	−2.854 21 (−2.847 656 2)[①]	−7.229 29	−13.604 3	−21.979 4	−32.354 4	−44.729 4	−59.104 4	−75.479 4
Total kinetic energy (K.E.)	2.854 21 (2.847 656 2)[①]	7.229 29	13.604 3	21.979 4	32.354 4	44.729 4	59.104 4	75.479 4
Total nuclear attractive energy	−6.750 28	−16.125 16	−29.500 2	−46.875 0	−68.250 0	−93.625 0	−123.000 0	−156.375 0
Electron repulsive energy	1.041 86	1.666 57	2.291 45	2.916 38	3.541 33	4.166 3	4.791 27	5.416 26
Total electron potential energy (P.E.)	−5.708 42 (−5.695 312 5)[①]	−14.458 6	−27.208 7	−43.958 7	−64.708 7	−89.458 8	−118.209	−150.959
P.E./K.E.	−2	−2	−2	−2	−2	−2	−2	−2

Notes ① The values in the parenthesis are Hartree–Fock (HF) values, and are taken from Clementi E, Roetti C. Roothaan–Hartree–Fock atomic wavefunction: basis functions and their coefficients for ground and certain excited states of neutral and ionized atoms, $Z \leq 54$. At Data Nucl Data, 1974, 14:445

If we further let $d = 0$ and write Z' as ζ (Zeta), Eq. (4.4.28) becomes

$$W = \xi^2 - 2Z\xi + \frac{5}{8}\xi \tag{4.4.29}$$

The formula has a minimum when $\zeta = Z - \frac{5}{16}$, i.e. $W_{\min} = -(Z - \frac{5}{16})^2$. This restores the variational results by using the effective nuclear charge ζ to replace atom number Z in the hydrogen-like wave function.

If $d = 0$, $Z' = Z$, Eq. (4.4.28) becomes

$$W = -Z^2 + \frac{5}{8}Z \tag{4.4.30}$$

This then restores the results of He-like-sequence treated by first-order perturbation of the wave function of zero order which is hydrogen-like wave function.

Based on the above description and derivation, the following meaningful conclusions can be drawn:

(1) $\psi_I(1)$ and $\psi_{II}(2)$ in Eq. (4.4.4) have different forms which indicates that we didn't assume $\psi_I(1)$ and $\psi_{II}(2)$ are the same beforehand, but when searching for the minimum of W, it naturally occurs that when $Z'_1 = Z'_2$ and $d_1 = d_2$, W has the minimum. $Z'_1 = Z'_2$ and $d_1 = d_2$ mean that both the electron that is newly added into the system of He$^+$-sequence and the old electron are on the 1s orbital, and they have the same spatial wave functions of the spin orbital as well as opposite and indistinguishable spins. Therefore, the way of WBEPM Theory satisfies Pauli exclusion principle and lowest energy theorem.
(2) The treatment of He ground state using WBEPM Theory can restore the results of ground-state He atom by variational method and perturbation method in quantum chemistry.
(3) The results of calculation, P.E./K.E. $= -2$, satisfies Virial theorem.

All the above three points show that the mode of adding electrons in WBEPM Theory that is stated in Chap. 2 accords with the basic theorem of quantum mechanics.

Table 4.46 shows that the results of He-sequence treated by the method of single generalized Laguerre polynomial is a little bit better than the results of Hartree–Fock (HF) method using single ζ (Zeta) function, but still have certain deviation from the experimental value. Let's take helium atom as example. The deviation is due to neglecting relativistic effect, which brings a deviation of -0.00007 Hartree [168], and the use of a single determinant causes the lack of consideration of electron correlation, which will introduce a correlation energy deviation of -0.04204 Hartree [169]. The remaining deviation of about -0.0074 Hartree comes from the use of single generalized Laguerre polynomial as the trial function (the experimental value of the total electron energy of He atom is equal to -2.90372 Hartree [170]). In the next section, readers will see that this deviation will be largely reduced after replacing single generalized Laguerre function with double generalized Laguerre function.

4.4 Calculation of Total Electron Energy

Other valuable results (which are listed in Table 4.47) are obtained during the treatment of He-sequence using WBEPM Theory. Below, the origins of these results will be explained.

(1) Orbital energy
   In the self-consistent field method, there is

$$E = \frac{1}{2}\sum_i (\epsilon_i + f_i) \qquad (4.4.31)$$

where ϵ_i is the orbital energy, E is the total electron energy, and f_i is the sum of kinetic energy and nuclear attractive energy. If substituting the corresponding data in Table 4.46 into Eq. (4.4.31), the orbital energy can be obtained under the idea of self-consistence field method, and take He as example, $E = -2.85421$ a.u., $\sum_i f_i = 2.85421 + (-6.75028) = -3.89607$ a.u., then we get $\epsilon_i = -0.90618$ a.u.

(2) Gain of Koopman's theorem
   Use Δ to represent gain, then

$$\Delta = I_k - (E^+ - E) \qquad (4.4.32)$$

where I_k represents the ionization energy under Koopman's theorem, E^+ and E represent the total electron energy of the system before and after ionization, respectively. Take He as example, using data in Table 4.46

$$I_k = -\left[\begin{array}{l}\frac{1}{2}(\text{total kinetic energy} + \text{total nuclear attractive energy}) \\ + \text{electron repulsive energy}\end{array}\right]$$

$$= -\left[\frac{1}{2}(2.85421 - 6.75028) + 1.04186\right] = 0.90618 \,(\text{a.u.})$$

while $E = -2.85421$ a.u., $E^+ = -I_{exp} = -54.417760$ eV.
We get

$$\Delta = 0.05197 \text{ a.u.} = 1.4142 \text{ eV}$$

(3) Relaxation effect
   The change of the energy which is correlated with $\left(-\frac{1}{2}\nabla^2 - \frac{Z}{r}\right)$ caused by adding electron, taking He as example, is equal to $\left(-\frac{54.417760}{27.2116}\right) - \frac{1}{2}[(2.85421 - 6.75028)] = -0.052$ (a.u.) $= -1.414$ (eV) using data in Table 4.46. Note the relation between relaxation energy and gain value.

(4) Slater screening constant
   The electron energy of Slater-type function is $\varepsilon = -\frac{(Z-s)^2}{2n^{*2}}$. Let's still take He as example, substituting the data in Table 4.46 into it, then

Table 4.47 Some other valuable results

	He I	Li II	Be III	B IV	C V	N VI	O VII	F VII
Z'	1.539 29	2.533 4	3.530 61	4.528 98	5.527 91	6.527 15	7.526 59	8.526 16
n'	0.955 057	0.970 903	0.978 493	0.982 945	0.985 87	0.987 938	0.989 479	0.990 671
Orbital energy a.u.	−0.906 18 (−0.896 48)①	−2.781 36	−5.656 37	−9.5316	−14.406 6	−20.281 6	−27.156 6	−35.031 6
Gain of Koopman's theorem a.u.	0.051 97	0.052 07	0.052 18	0.052 02	0.052 07	0.052 1	0.052 13	0.052 14
Relaxation energy a.u.	−0.052	0.052 07	−0.052 1	−0.052 2	−0.052 2	−0.052 2	−0.052 2	−0.052 2
Screening constant②	0.310 559 21 (0.3)	0.311 266 (0.3)	0.311 599 2 (0.3)	0.311 780 7 (0.3)	0.311 907 1 (0.3)	0.311 995 8 (0.3)	0.312 061 3 (0.3)	0.312 111 8 (0.3)
Radial expectation value	0.902 8	0.563 7	0.409 8	0.3219	0.2650	0.2225 2	0.195 8	0.173 2

Notes ① The orbital energy in the parenthesis is the single ζ HF value, and is taken from Clementi E. At Data Nucl Data Tables, 1974, 14:445
② The screening constant in the parenthesis is the value given by Slater, and is taken from Slater JC. Phys Rev, 1930, 36:57

4.4 Calculation of Total Electron Energy

$$\frac{1}{2}E = \frac{1}{2} \times (-2.85421) = -\frac{(2-s)^2}{2 \times 1^2}$$

We get

$$s = 0.3105592$$

(5) Radial expectation value
Previously we have derived

$$\langle r \rangle = \frac{3n'^2 - l'(l'+1)}{2Z'} \qquad (4.4.33)$$

Take He as example, by substituting the data in Table 4.46 into it, we get

$$\langle r \rangle = 0.9028$$

4.4.2.2 Double Generalized Laguerre Function as a Trial Function [160]

In order to increase the accuracy of calculation and at the same time indicate that the linear combination technique that is widely used in the calculation of quantum chemistry can completely apply to the calculation of WBEPM Theory, in this section, we will use the one-electron wave function of the weakest bound electron in the form of a linear combination of double generalized Laguerre functions to replace the one-electron wave function of the weakest bound electron expressed by a single generalized Laguerre function as a trial function for the variational treatment of He atom.

In the last section we have shown that when the spatial functions of two electrons are identical, the approximate total electron energy W of the He-sequence ground state has minimum. Thus, this result will be used for the treatment in this section, i.e. the total electron wave function of ground-state He atom is taken as

$$\Psi = (2!)^{-\frac{1}{2}} \sum_p (-1)^p P\{\psi_{1s}(1)\alpha(1)\psi_{1s}(2)\beta(2)\} \qquad (4.4.34)$$

or

$$\Psi = \left| \psi_{1s}(1)\overline{\psi_{1s}(2)} \right| \qquad (4.4.35)$$

where

$$\psi_{1s}(\mu) = R(\mu)Y_{00}(\theta_\mu, \phi_\mu) = c_1\varphi_1(\mu) + c_2\varphi_2(\mu) \qquad (4.4.36)$$

In Eq. (4.4.36),

$$\varphi_i(\mu) = \frac{1}{\sqrt{4\pi \Gamma(2l_i' + 3)}} \left(\frac{2Z_i'}{n_i'}\right)^{l_i' + \frac{3}{2}} r_\mu^{l_i'} e^{-\frac{Z_i' r_\mu}{n_i'}} \tag{4.4.37}$$

Here, $n_i' = n_i + d_i$, $l_i' = l_i + d_i$, n_i and l_i are principal quantum number and azimuthal quantum number, respectively. Z_i and d_i are undetermined parameters.

The expectation value of the approximate total electron energy of the ground-state He atom is

$$W = \frac{\langle \Psi | \hat{H} | \Psi \rangle}{\langle \Psi | \Psi \rangle} \tag{4.4.38}$$

where

$$\hat{H} = -\frac{1}{2}\nabla_1^2 - \frac{1}{2}\nabla_2^2 - \frac{Z}{r_1} - \frac{Z}{r_2} + \frac{1}{r_{12}} = \sum_{\mu=1}^2 \hat{h}(\mu) + 1/r_{12} \tag{4.4.39}$$

We can derive

$$\langle \Psi | \Psi \rangle = \left[(c_1^2 + c_2^2) + 2c_1 c_2 S\right]^2 \tag{4.4.40}$$

and

$$\langle \Psi | \hat{H} | \Psi \rangle = c_1^4 A + 4c_1^3 c_2 B + 4c_1^2 c_2^2 C + 2c_1^2 c_2^2 D + 4c_1 c_2^3 E + c_2^4 F \tag{4.4.41}$$

A, B, C, D, E and F in Eq. (4.4.41), respectively, represents the following integrals

$$\begin{aligned} A &= \langle \varphi_1(1)\varphi_1(2) | \hat{H} | \varphi_1(1)\varphi_1(2) \rangle \\ &= \langle \varphi_1(1) | \hat{h}(1) | \varphi_1(1) \rangle + \langle \varphi_1(2) | \hat{h}(2) | \varphi_1(2) \rangle \\ &\quad + \langle \varphi_1(1)\varphi_1(2) | \frac{1}{r_{12}} | \varphi_1(1)\varphi_1(2) \rangle \end{aligned} \tag{4.4.42}$$

$$\begin{aligned} B &= \langle \varphi_1(1)\varphi_1(2) | \hat{H} | \varphi_2(1)\varphi_1(2) \rangle \\ &= \langle \varphi_1(1) | \hat{h}(1) | \varphi_2(1) \rangle + \langle \varphi_1(2) | \hat{h}(2) | \varphi_1(2) \rangle S \\ &\quad + \langle \varphi_1(1)\varphi_1(2) | \frac{1}{r_{12}} | \varphi_2(1)\varphi_1(2) \rangle \end{aligned} \tag{4.4.43}$$

4.4 Calculation of Total Electron Energy

$$C = \langle \varphi_1(1)\varphi_1(2) | \hat{H} | \varphi_2(1)\varphi_2(2) \rangle$$
$$= \langle \varphi_1(1) | \hat{h}(1) | \varphi_2(1) \rangle S + \langle \varphi_1(2) | \hat{h}(2) | \varphi_2(2) \rangle S$$
$$+ \left\langle \varphi_1(1)\varphi_1(2) \left| \frac{1}{r_{12}} \right| \varphi_2(1)\varphi_2(2) \right\rangle \tag{4.4.44}$$

$$D = \langle \varphi_2(1)\varphi_1(2) | \hat{H} | \varphi_2(1)\varphi_1(2) \rangle$$
$$= \langle \varphi_2(1) | \hat{h}(1) | \varphi_2(1) \rangle + \langle \varphi_1(2) | \hat{h}(2) | \varphi_1(2) \rangle$$
$$+ \left\langle \varphi_2(1)\varphi_1(2) \left| \frac{1}{r_{12}} \right| \varphi_2(1)\varphi_1(2) \right\rangle \tag{4.4.45}$$

$$E = \langle \varphi_2(1)\varphi_1(2) | \hat{H} | \varphi_2(1)\varphi_2(2) \rangle$$
$$= \langle \varphi_2(1) | \hat{h}(1) | \varphi_2(1) \rangle S + \langle \varphi_1(2) | \hat{h}(2) | \varphi_2(2) \rangle$$
$$+ \left\langle \varphi_2(1)\varphi_1(2) \left| \frac{1}{r_{12}} \right| \varphi_2(1)\varphi_2(2) \right\rangle \tag{4.4.46}$$

$$F = \langle \varphi_2(1)\varphi_2(2) | \hat{H} | \varphi_2(1)\varphi_2(2) \rangle$$
$$= \langle \varphi_2(1) | \hat{h}(1) | \varphi_2(1) \rangle + \langle \varphi_2(2) | \hat{h}(2) | \varphi_2(2) \rangle$$
$$+ \left\langle \varphi_2(1)\varphi_2(2) \left| \frac{1}{r_{12}} \right| \varphi_2(1)\varphi_2(2) \right\rangle \tag{4.4.47}$$

In Eqs. (4.4.40), (4.4.43), (4.4.44), and (4.4.46)

$$S = \langle \varphi_1(1)\varphi_2(1) \rangle$$
$$= \left[\Gamma(2l'_1 + 3)\Gamma(2l'_2 + 3) \right]^{-1/2} \left(\frac{2Z'_1}{n'_1} \right)^{l'_1 + 3/2}$$
$$\times \left(\frac{2Z'_2}{n'_2} \right)^{l'_2 + 3/2} \left(\frac{Z'_1}{n'_1} + \frac{Z'_2}{n'_2} \right)^{-l'_1 - l'_2 - 3} \Gamma(l'_1 + l'_2 + 3) \tag{4.4.48}$$

The formulae for all integrals in Eq. (4.4.42) to Eq. (4.4.47) have been derived, and the specific formulae are as follows:

$$\langle \varphi_i(\mu) | \hat{h}(\mu) | \varphi_i(\mu) \rangle = \left\langle \varphi_i(\mu) \left| -\frac{1}{2}\nabla_\mu^2 - \frac{Z}{r_\mu} \right| \varphi_i(\mu) \right\rangle$$
$$= \frac{Z'^2_i}{2n'^2_i(2n'_i + 1)} - \frac{ZZ'_i}{n'^2} \tag{4.4.49}$$

$$\langle \varphi_i(\mu) | \hat{h}(\mu) | \varphi_j(\mu) \rangle$$

$$= \left\langle \varphi_i(\mu) \left| -\frac{1}{2}\nabla_\mu^2 - \frac{Z}{r_\mu} \right| \varphi_j(\mu) \right\rangle$$

$$= [\Gamma(2l'_i + 3)\Gamma(2l'_j + 3)]^{-\frac{1}{2}} \left(\frac{2Z'_i}{n'_i}\right)^{l'_i+\frac{3}{2}} \left(\frac{2Z'_j}{n'_j}\right)^{l'_j+\frac{3}{2}}$$

$$\times \left(\frac{Z'_i}{n'_i} + \frac{Z'_j}{n'_j}\right)^{-l'_i-l'_j-1} \Gamma(l'_i + l'_j + 1)$$

$$\times \left[-\frac{Z'^2_j}{2n'^2_j}(l'_i + l'_j + 2)(l'_i + l'_j + 1)\left(\frac{Z'_i}{n'_i} + \frac{Z'_j}{n'_j}\right)^{-2} \right.$$

$$\left. + Z'_j(l'_i + l'_j + 1)\left(\frac{Z'_i}{n'_i} + \frac{Z'_j}{n'_j}\right)^{-1} - \frac{l'_j(l'_j + 1)}{2} \right]$$

$$- Z[\Gamma(2l'_i + 3)\Gamma(2l'_j + 3)]^{-\frac{1}{2}} \left(\frac{2Z'_i}{n'_i}\right)^{l'_i+\frac{3}{2}} \left(\frac{2Z'_j}{n'_j}\right)^{l'_j+\frac{3}{2}}$$

$$\times \left(\frac{Z'_i}{n'_i} + \frac{Z'_j}{n'_j}\right)^{-l'_i-l'_j-2} \Gamma(l'_i + l'_j + 2) \quad (4.4.50)$$

$$\left\langle \varphi_i(\mu)\varphi_i(\nu) \frac{1}{r_{\mu\nu}} \varphi_i(\mu)\varphi_i(\nu) \right\rangle$$

$$= \frac{Z'_i}{n'^2_i} \left\{ 1 - 2^{-4n'_i} \Gamma(4n'_i + 1)[\Gamma(2n'_i + 1)]^{-2} \right\} \quad (4.4.51)$$

$$\left\langle \varphi_1(1)\varphi_1(2) \frac{1}{r_{12}} \varphi_2(1)\varphi_1(2) \right\rangle$$

$$= [\Gamma(2l'_1 + 3)\Gamma(2l'_2 + 3)]^{-\frac{1}{2}} \left(\frac{2Z'_2}{n'_2}\right)^{l'_2+\frac{3}{2}}$$

$$\times \left\{ \left(\frac{2Z'_1}{n'_1}\right)^{l'_1+\frac{5}{2}} \left(\frac{Z'_1}{n'_1} + \frac{Z'_2}{n'_2}\right)^{l'_1-l'_2-3} (2l'_1 + 2)^{-1} \Gamma(l'_1 + l'_2 + 3) \right.$$

$$- \left(\frac{2Z'_1}{n'_1}\right)^{-l'_2-\frac{1}{2}} \left[(2Z'_1 n'_2)^{-1}(3Z'_1 n'_2 + Z'_2 n'_1)\right]^{-3l'_1-l'_2-4}$$

$$\times [\Gamma(2l'_1 + 3)]^{-1} \Gamma(3l'_1 + l'_2 + 4) + \left(\frac{2Z'_1}{n'_1}\right)^{-l'_2-\frac{1}{2}}$$

$$\left. \times [\Gamma(2l'_1 + 3)]^{-1} \left\{ \Gamma(2l'_1 + 3)\left[(2Z'_1 n'_2)^{-1}(Z'_1 n'_2 + Z'_2 n'_1)\right]^{-l'_1-l'_2-2} \right. \right.$$

4.4 Calculation of Total Electron Energy

$$\times \Gamma(l_1' + l_2' + 2) - \Gamma(2l_1' + 2)$$

$$\times \left[(2Z_1'n_2')^{-1}(Z_1'n_2' + Z_2'n_1')\right]^{-l_1'-l_2'-3} \Gamma(l_1' + l_2' + 3)$$

$$\int_0^\infty [(2l_1' + 2) - x] x_1^{l_1'+l_2'+1} \exp\left(\frac{-Z_1'n_2' + Z_2'n_1'x}{2Z_1'n_2'}\right)$$

$$\Gamma(2l_1' + 2, x) \mathrm{d}x \}\tag{4.4.52}$$

$$\left\langle \varphi_1(1)\varphi_1(2) \frac{1}{r_{12}} \varphi_2(1)\varphi_2(2) \right\rangle$$

$$= \left[\Gamma(2l_1' + 3)\Gamma(2l_2' + 3)\right]^{-\frac{1}{2}} \left(\frac{2Z_1'}{n_1'}\right)^{2l_1'+3} \left(\frac{2Z_2'}{n_2'}\right)^{2l_2'+3}$$

$$\times \left(\frac{Z_1'}{n_1'} + \frac{Z_2'}{n_2'}\right)^{-2l_1'-2l_2'-5} \left[\Gamma(l_1' + l_2' + 2)\Gamma(l_1' + l_2' + 3)\right.$$

$$\left. - 2^{-2l_1'-2l_2'-3}\Gamma(2l_1' + 2l_2' + 4)\right]\tag{4.4.53}$$

$$\left\langle \varphi_2(1)\varphi_1(2) \frac{1}{r_{12}} \varphi_2(1)\varphi_1(2) \right\rangle$$

$$= \frac{Z_2'}{n_2'^2} - \left[\Gamma(2l_1' + 3)\Gamma(2l_2' + 3)\right]^{-1} \left(\frac{2Z_2'}{n_2'}\right)^{2l_2'+3}$$

$$\times \left\{ \left(\frac{2Z_1'}{n_1'}\right)^{2l_1'+2} \left(\frac{2Z_1'}{n_1'} + \frac{2Z_2'}{n_2'}\right)^{-2l_1'-2l_2'-4} \Gamma(2l_1' + 2l_2' + 4) \right.$$

$$\left. - \left(\frac{2Z_1'}{n_1'}\right)^{-2l_2'-2} \int_0^\infty [(2l_1' + 2) - x] x^{2l_2'+1} \right.$$

$$\left. \exp\left(\frac{-Z_2'n_1'x}{Z_1'n_2'}\right) \Gamma(2l_1' + 2, x) \mathrm{d}x \right\}\tag{4.4.54}$$

$$\langle \varphi_2(1)\varphi_1(2) 1/r_{12} \varphi_2(1)\varphi_2(2) \rangle$$
$$= \left[\Gamma(2l_1' + 3)\Gamma(2l_2' + 3)\right]^{-1/2} ((2Z_1')/(n_1'))^{l_1'+3/2} ((2Z_2')/(n_2'))^{l_2'+3/2}$$
$$\times ((Z_1')/(n_1') + (Z_2')/(n_2'))^{-l_1'-l_2'-2} \Gamma(l_1' + l_2' + 2) + \left[\Gamma(2l_1' + 3)\right]^{-1/2}$$
$$\times \left[\Gamma(2l_2' + 3)\right]^{-3/2} ((2Z_1')/(n_1'))^{l_1'+3/2}$$
$$((2Z_2')/(n_2'))^{3l_2'+9/2} ((Z_1')/(n_1') + (Z_2')/(n_2'))^{-l_1'-3l_2'-5}$$
$$\times \left\{ -\left[(Z_1'n_2' + Z_2'n_1')^{-1}(Z_1'n_2' + 3Z_2'n_1')\right]^{-l_1'-3l_2'-4} \right.$$
$$\times \Gamma(l_1' + 3l_2' + 4) + (l_1' + l_2' + 2)\Gamma(l_1' + l_2' + 2)$$

$$\times \Gamma(2l_2' + 2) \left[(Z_1'n_2' + Z_2'n_1')^{-1} (2Z_2'n_1') \right]^{-2l_2'-2} - \Gamma(l_1' + l_2' + 2)$$

$$\times \Gamma(2l_2' + 3) \left[(Z_1'n_2' + Z_2'n_1')^{-1} (2Z_2'n_1') \right]^{-2l_2'-3}$$

$$\int_0^\infty \left[(l_1' + l_2' + 2) - x \right] x^{2l_2'+1} exp\left((-2Z_2'n_1'x)/(Z_1'n_2' + Z_2'n_1') \right)$$

$$\times \Gamma(l_1' + l_2' + 2, x) dx \} \qquad (4.4.55)$$

By substituting related formulae in Eq. (4.4.41) to Eq. (4.4.55) into Eq. (4.4.38), we get the formula for W.

By searching for the minimum of W, when $c_1 = 0.843$, $Z_1' = 1.45328$, $d_1 = -0.000081$, $c_2 = 0.1814$, $Z_2' = 2.9$ and $d_2 = -0.00086$, $W_{min} = -2.861672864$ a.u.

Because we are looking for the minimum, it is hard to find an exact minimum. In fact, several other minimums were found around the above given $W_{min} = -2.861672864$ a.u., and related parameters and minimums are listed in Table 4.48.

The comparison of the results obtained by using one-electron wave function of the weakest bound electron, which is in the form of a linear combination of double generalized Laguerre functions, as a trial function with other results are listed in Table 4.49.

From Table 4.49 we can see that: ① the results calculated by using a linear combination of double generalized Laguerre functions within WBEPM Theory is a little bit better than the results of Hartree–Fock (HF) method using double ζ (Zeta) functions, and if $d_i = 0$, i.e. $n_i' = n_i + d_i = n_i$, $l_i' = l_i + d_i = l_i$, Eq. (4.4.37) will be reverted to Slater function so that the results of HF method using double ζ (Zeta)

Table 4.48 Several minimums and related parameters of W that were found

Serial No.	1	2	3	4①	5
Total electron energy (T.E.)	−2.861 672 955	−2.861 672 971	−2.861 672 860	−2.861 672 864	−2.861 672 846
c_1	0.846 94	0.846 928	0.843	0.843	0.843
c_2	0.184 04	0.184 127	0.181 4	0.181 4	0.181 4
Z_1'	1.451 94	1.451 96	1.453 22	1.453 28	1.453 16
Z_2'	1.893 81	2.892 72	2.899 86	2.9	2.899 72
d_1	−0.000 081	−0.000 058	−0.000 081	−0.000 081	−0.000 081
d_2	−0.000 86	−0.001	−0.000 86	−0.000 86	−0.000 86
P.E. / K.E. ②	−1.999 997 8	−1.999 987 0	−2.000 067 92	−1.999 981 77	2.000 154 08
$\psi(\mu)$ normalization	1.012 729 54	1.012 877 89	1.000 000 76	0.999 999 87	1.000 001 66

Notes ① The value for number 4 is the one given in the main text, $W_{min} = -2.861672864$ a.u. ≈ -2.8616729 a.u.
② P.E./K.E. is the ratio of potential energy to kinetic energy, and these ratios are all close to -2, indicating Virial theorem is satisfied

4.4 Calculation of Total Electron Energy

Table 4.49 The comparison of the calculated values by a linear combination of double generalized Laguerre functions with other related values[①]

Serial No.	1	2	3	4
Total electron energy (T.E>) of ground-state He atom (T.E.)	−2.861 672 9	−2.861 672 6	−2.854 21	−2.847 656 2
Total kinetic energy (K.E.)	2.861 725 0	2.861 685 5	2.854 21	2.847 656 2
Total potential energy (P.E.)	−5.723 397 9	−5.723 358 1	−5.708 42	−5.695 312 5
P.E./T.E.	−1.999 981 8	−1.999 995 5	−2	−2.000 000 0
Total repulsive potential energy	1.025 872 0		1.041 86	
Total attractive potential energy	−6.749 269 9		−6.750 28	
$\psi(\mu)$ normalization	0.999 999 9	0.999 995 6	1	1
Orbital energy	−0.917 900 4	−0.917 94	−0.906 18	−0.896 48

Notes ① In the table: 1 represents the calculated results using double generalized Laguerre functions within WBEPM Theory
2 represents the results of Hartree–Fock (HF) method using double ζ (Zeta) functions. The results are taken from Clementi E, Roetti C. At Data Nucl Data, 1974, 14:428
3 represents the calculated results using a single generalized Laguerre function within WBEPM Theory
4 represents the results of HF method using a single (Zeta) function. The source of the results is the same as for 2

functions will also be obtained; ② comparing with the results calculated by using a single generalized Laguerre function within WBEPM Theory, there is large improvement in the results calculated by using a linear combination of double generalized Laguerre functions within WBEPM Theory, which indicates that our expectation is accomplished, i.e. increasing the accuracy of the calculation, and which at the same time also indicates that the linear combination technique in which atomic orbitals or molecular orbitals can be represented by a set of appropriate basis functions works the same way in WBEPM Theory. More generally, any appropriate set of basis functions, including Slater function, Gaussian function, generalized Laguerre function, hydrogen-like wave function, etc. [171, 172], is applicable in WBEPM Theory. ③ The result by using a linear combination of double generalized Laguerre functions is −2.861 672 9 a.u., and the experimental value is −2.903 72 a.u. [170]. If we add the calculated value and relativistic correction (−0.000 07 a.u.) as well as correction of correlation energy (−0.042 04 a.u.) together, the result is almost the same as the experimental value. This indicates that the calculation error is due to the use of non-relativistic Hamiltonian and the use of single-determinant antisymmetry wave functions, which results in the neglect of relativistic effect as well as biased estimation of electron correlation effect. From previous theoretical elaboration, we have shown that there is room in WBEPM Theory to accommodate the corrections for these effects, and all effective theoretical methods for the calculation of relativity and correlation energy in existing quantum chemistry methods can be introduced into the theory of WBEPM Thoery.

4.4.3 Perturbation Treatment on the Energy of the He-Sequence Ground State with the WBEPM Theory [160]

For the He-sequence ground state, the non-relativistic Hamiltonian of the system is

$$\hat{H} = \frac{1}{2}\nabla_1^2 - \frac{1}{2}\nabla_2^2 - \frac{Z}{r_1} - \frac{Z}{r_2} + \frac{1}{r_{12}} = \sum_{i=1}^{2} \hat{H}_i \quad (4.4.56)$$

$\hat{H}_i$ can be divided into two parts:

$$\hat{H}_i = \left[-\frac{1}{2}\nabla_i^2 + V(r_i) \right] + \left[-V(r_i) - \frac{Z}{r_i} + \sum_{j=i+1}^{2} \frac{1}{r_{ij}} \right] = H_i^0 + H_i' \quad (4.4.57)$$

Here, H_1^0 represents unperturbed Hamiltonian and H_i' represents perturbation. Substituting Eq. (4.4.57) into Eq. (4.4.56), one has

$$\hat{H} = \sum_{i=1}^{2} H_i^0 + \sum_{i=1}^{2} H_i' = H^0 + H' \quad (4.4.58)$$

Let

$$V(r_i) = -\frac{Z_i'}{r_i} + \frac{d_i(d_i+1) + 2d_i l_i}{2r_i^2} \quad (4.4.59)$$

then

$$H^0 = H_1^0 + H_2^0 = -\frac{1}{2}\nabla_1^2 - \frac{Z_1'}{r_1} + \frac{d_1(d_1+1) + 2d_1 l_1}{2r_1^2}$$
$$- \frac{1}{2}\nabla_2^2 - \frac{Z_2'}{r_2} + \frac{d_2(d_2+1) + 2d_2 l_2}{2r_2^2} \quad (4.4.60)$$

As stated in Chap. 3, H^0 is the Hamiltonian of the problem that can be solved exactly. The corresponding unperturbed wave function is

$$\Psi^0(r_1, r_2) = \psi_{1s}^2(r_1)\psi_{1s}^2(r_2)$$
$$= \frac{1}{4\pi\sqrt{\Gamma(2l_1'+3)\Gamma(2l_2'+3)}} \left(\frac{2Z_1'}{n_1'}\right)^{l_1'+3/2}$$
$$\times \left(\frac{2Z_2'}{n_2'}\right)^{l_2'+\frac{3}{2}} r_1^{l_1'} r_2^{l_2'} e^{-(Z_1' r_1/n_1') - (Z_2' r_2/n_2')} \quad (4.4.61)$$

The total electron energy without perturbation is

4.4 Calculation of Total Electron Energy

$$E^0 = \varepsilon_1^0 + \varepsilon_2^0 = -\frac{Z_1'^2}{2n_1'^2} - \frac{Z_2'^2}{2n_2'^2} \qquad (4.4.62)$$

Now let's calculate the energy of first order perturbation

$$\begin{aligned}
E' &= \langle \Psi^0(r_1, r_2) | H' | \Psi^0(r_1, r_2) \rangle \\
&= \langle \Psi^0(r_1, r_2) | H_1' | \Psi^0(r_1, r_2) \rangle \\
&\quad + \langle \Psi^0(r_1, r_2) | H_2' | \Psi^0(r_1, r_2) \rangle \\
&= \left[\left\langle \psi_{1s}^0(r_1) \left| \frac{Z_1' - Z}{r_1} - \frac{d_1(d_1+1) + 2d_1 l_1}{2r_1^2} \right| \psi_{1s}^0(r_1) \right\rangle \right. \\
&\quad + \left\langle \psi_{1s}^0(r_1) \psi_{1s}^0(r_2) \left| \frac{1}{r_{12}} \right| \psi_{1s}^0(r_1) \psi_{1s}^0(r_2) \right\rangle \right] \\
&\quad + \left[\left\langle \psi_{1s}^0(r_2) \left| \frac{Z_2' - Z}{r_2} - \frac{d_2(d_2+1) + 2d_2 l_2}{2r_2^2} \right| \psi_{1s}^0(r_2) \right\rangle \right] \\
&= E_1' + E_2' \qquad (4.4.63)
\end{aligned}$$

According to Ref. [173], there is

$$\left\langle \psi_{1s} \left| \frac{1}{r_1} \right| \psi_{1s} \right\rangle = \frac{Z'}{n'^2} \qquad (4.4.64)$$

and

$$\left\langle \psi_{1s} \left| \frac{1}{r^2} \right| \psi_{1s} \right\rangle = \frac{2Z'^2}{n'^3 (2l'+1)} \qquad (4.4.65)$$

The expansion of $\frac{1}{r_{ij}}$ in spherical coordinates is [86]

$$\frac{1}{r_{ij}} = \frac{1}{r_>} \sum_0^\infty \frac{r_<^l}{r_>^l} P_l(\cos\theta) \qquad (4.4.66)$$

Then we can derive

$$\begin{aligned}
&\left\langle \psi_{1s}^0(r_1) \psi_{1s}^0(r_2) \left| \frac{1}{r_{12}} \right| \psi_{1s}^0(r_1) \psi_{1s}^0(r_2) \right\rangle \\
&= \frac{Z_1'}{n_1'^2} - \frac{1}{\Gamma(2l_1'+3)\Gamma(2l_2'+3)} \left(\frac{2Z_1'}{n_1'}\right)^{2l_1'+3} \left(\frac{2Z_2'}{n_2'}\right)^{2l_2'+2} \\
&\quad \times \left(\frac{2Z_1'}{n_1'} + \frac{2Z_2'}{n_2'}\right)^{-(2l_1'+2l_2'+4)} \Gamma(2l_1'+2l_2'+4) \\
&\quad - \frac{1}{\Gamma(2l_1'+3)\Gamma(2l_2'+3)} \left(\frac{2Z_1'}{n_1'}\right)^{2l_1'+3} \left(\frac{2Z_2'}{n_2'}\right)^{-2l_2'-2}
\end{aligned}$$

$$\times \left\{ \int_0^\infty \left[(2l_2' + 2)r^{2l_1'+1} - r^{2l_1'+2} \right] \exp\left(\frac{-Z_1' n_2' r}{Z_2' n_1'} \right) \right.$$
$$\left. \times \Gamma(2l_2' + 2, x) dx \right\} \tag{4.4.67}$$

So

$$E_1' = \langle \psi_{1s}^0(r_1) | H_1' | \psi_{1s}^0(r_1) \rangle$$
$$= \frac{Z_1'^2}{n_1'^2} \left(1 - \frac{l_1'}{2l_1' + 1} \right) - \frac{ZZ_1'}{n_1'^2}$$
$$+ \langle \psi_{1s}^0(r_1) \psi_{1s}^0(r_2) | \frac{1}{r_{12}} | \psi_{1s}^0(r_1) \psi_{1s}^0(r_2) \rangle \tag{4.4.68}$$

$$E_2' = \langle \psi_{1s}^0(r_2) | H_2' | \psi_{1s}^0(r_2) \rangle$$
$$= \frac{Z_2'^2}{n_2'^2} \left(1 - \frac{l_2'}{2l_2' + 1} \right) - \frac{ZZ_2'}{n_2'^2} \tag{4.4.69}$$

The correction for the energy of the first order perturbation is

$$E' = E_1' + E_2' \tag{4.4.70}$$

Adding this correction for the energy of the first order perturbation to the unperturbed energy, we get

$$E = E^0 + E^1$$
$$= \frac{Z_1'^2}{2n_1'^2(2l_1' + 1)} - \frac{ZZ_1'}{n_1'^2} + \frac{Z_2'^2}{2n_2'^2(2l_2' + 1)} - \frac{ZZ_2'}{n_2'^2}$$
$$+ \frac{Z_1'}{n_1'^2} - \frac{1}{\Gamma(2l_1' + 3)\Gamma(2l_2' + 3)} \left(\frac{2Z_1'}{n_1'} \right)^{2l_1'+3} \left(\frac{2Z_2'}{n_2'} \right)^{2l_2'+3}$$
$$\times \left(\frac{2Z_1'}{n_1'} + \frac{2Z_2'}{n_2'} \right)^{-(2l_1'+2l_2'+4)} \Gamma(2l_1' + 2l_2' + 4)$$
$$- \frac{1}{\Gamma(2l_1' + 3)\Gamma(2l_2' + 3)} \left(\frac{2Z_1'}{n_1'} \right)^{2l_1'+3} \left(\frac{2Z_2'}{n_2'} \right)^{-2l_2'-2}$$
$$\times \left\{ \int_0^\infty \left[(2l_2' + 2)r^{2l_1'+1} - r^{2l_1'+2} \right] \exp\left(\frac{-Z_1' n_2' r}{Z_2' n_1'} \right) \right.$$
$$\left. \times \Gamma(2l_2' + 2, x) dx \right\} \tag{4.4.71}$$

By searching for the minimum total energy, the following result for He atom is obtained: when

$$\begin{cases} Z'_1 = Z'_2 = 1.53929 \\ d_1 = d_2 = -0.044926 \end{cases} \quad (4.4.72)$$

$$\min\left(E[Z_1^1, Z_2^1, d_1, d_2]\right) = -2.85421 \quad (4.4.73)$$

This result is the same as the result from variational treatment of the He-sequence using a single generalized Laguerre function.

4.5 Electronegativity, Hard and Soft Acids and Bases, and the Molecular Design of Coordination Polymers

4.5.1 The Electronegativity Concept and Scale

L. Pauling gave the definition of chemical bond in the book *The Nature of the Chemical Bond*: "We shall say that there is a chemical bond between two atoms or groups of atoms in case that the forces acting between them are such as to lead to the formation of an aggregate with sufficient stability to make it convenient for the chemist to consider it as an independent molecular species" [174]. Based on this definition, ionic bond, covalent bond, coordination bond, hydrogen bond and even contributions of Van der Waals force under some particular conditions (such as in the supramolecular compound and coordination polymer) all belong to the category of chemical bond.

The electronegativity scale is used to measure the ability of atom to attract electrons in a molecule [175]. So electronegativity is an important concept to understand chemical bond and molecular structure. Now, electronegativity has been connected with chemical potential, Lewis acid, base hardness, transition between chemical bond types and crystal structure, physical properties (such as bond energy, enthalpy of formation, etc.) and reactivity, molecular design and synthesis, etc.

The electronegativity is not a physical quantity and can't be measured by experiments. Thus, the size of electronegativity is characterized by scale, and Pauling first gave the thermochemical scale of electronegativity. Table 4.50 lists the values of Pauling electronegativity [176]. Besides Pauling electronegativity, many different types of electronegativity scale were proposed from various angles. There are Mulliken scale [177], Allred-Rochow scale [178], Gordy scale [179], Sanderson scale [180], Phillips scale [181], John-Bloch scale [182], Allen scale [183], Y. H. Zhang electronegativity scale of elements in valence states [184] and N. W. Zheng–G. S. Li nuclear potential scale [185], and Liu [186], Sun [187], Li [188], Gao [189], Yuan [190], et al. have also widely and deeply studied electronegativity scale. The concept of electronegativity is also connected to chemical potential [191, 192], and extended to orbital electronegativity, group electronegativity, etc. [193–199], from old electronegativity of the elements. This makes electronegativity have wide applications in chemistry, physics, material science, biology and geology.

Table 4.50 Pauling's electronegativity

H																	
2.1																	
Li	Be	B											C	N	O	F	
1.0	1.5	2.0											2.5	3.0	3.5	4.0	
Na	Mg	Al											Si	P	S	Cl	
0.9	1.2	1.5											1.8	2.1	2.5	3.0	
K	Ca	Sc	Ti	V	Cr	Mn	Fe	Co	Ni	Cu	Zn	Ga	Ge	As	Se	Br	
0.8	1.0	1.3	1.5	1.6	1.6	1.5	1.8	1.8	1.8	1.9	1.6	1.6	1.8	2.0	2.4	2.8	
Rb	Sr	Y	Zr	Nb	Mo	Tc	Ru	Rh	Pd	Ag	Cd	In	Sn	Sb	Te	I	
0.8	1.0	1.2	1.4	1.6	1.8	1.9	2.2	2.2	2.2	1.9	1.7	1.7	1.8	1.9	2.1	2.5	
Cs	Ba	La-Lu	Hf	Ta	W	Re	Os	Ir	Pt	Au	Hg	Tl	Pb	Bi	Po	At	
0.7	0.9	1.1 1.2	1.3	1.5	1.7	1.9	2.2	2.2	2.2	2.4	1.9	1.8	1.8	1.9	2.0	2.2	
Fr	Ra	Ac	Th	Pa	U	Np-No											
0.7	0.9	1.1	1.3	1.5	1.7	1.3											

Data source Pauling L. The nature of the chemical bond. Translated by Lu JX, Huang YC, Zeng GZ, et al. Shanghai Science and Technology Press, Shanghai, 1966

4.5.2 The Nuclear Potential Scale of the Weakest Bound Electron [185, 200]

Based on the description in Chap. 3, for the weakest bound electron, there are the following formulae:

$$\varepsilon_\mu^0 \approx -\frac{Z'^2}{2n'^2} \text{ (a. u.)} \tag{4.5.1}$$

$$I_\mu \approx -\varepsilon_\mu^0 \tag{4.5.2}$$

$$\langle r^{-1} \rangle = \frac{Z'}{n'^2} \tag{4.5.3}$$

$$\langle r \rangle = \frac{3n'^2 - l'(l'+1)}{2Z'} \tag{4.5.4}$$

In above formulae, ε_μ^0 represents the non-relativistic energy of the weakest bound electron in the central field, I_μ represents the ionization energy of the weakest bound electron, Z' is the effective nuclear charge that the weakest bound electron feels, n' and l' are called, respectively, the effective principle quantum number and the effective azimuthal quantum number, and $n' = n + d$, $l' = l + d$. $\langle r^{-1} \rangle$ and $\langle r \rangle$ are the expectation values for r^{-1} and r, respectively. Neglecting superscripts and

4.5 Electronegativity, Hard and Soft Acids and Bases …

subscripts, and combining equations from Eq. (4.5.1) to Eq. (4.5.3), we get

$$Z' = \frac{2I}{\langle r^{-1} \rangle} \tag{4.5.5}$$

Substituting Eq. (4.5.5) into Eq. (4.5.6), then

$$n' = \frac{(2I)^{\frac{1}{2}}}{\langle r^{-1} \rangle} \tag{4.5.6}$$

So Z' and n' can be represented by ionization energy I and expectation value $\langle r^{-1} \rangle$.

I can either take the experimental value of ionization energy or be calculated by the method given in Sect. 4.1, and $\langle r^{-1} \rangle$ and $\langle r \rangle$ can be calculated theoretically by quantum chemistry methods. Then the ratio Z'/r is obtained. We found that Z'/r has a good linear relationship with Mulliken electronegativity X_M ($X_M = (I + A)/2$) (see Fig. 4.3), and the linear correlation coefficient can be 0.971.

Mulliken electronegativity has three properties: ① $X_M = (I + A)/2$, which has intuitive physical meanings by using the average value of ionization energy and electron affinity to represent the capability to attract electron; ② there is a linear relation between Mulliken electronegativity and Pauling electronegativity [198]:

$$X_P = 0.366(X_M - 0.615) \tag{4.5.7}$$

③ Mulliken electronegativity is related to chemical potential, orbital electronegativity and hardness, thus it has profound theoretical foundations of quantum chemistry.

The good linear relation between the ratio $Z'/\langle r \rangle$ and Mulliken electronegativity indicates that the ratio $Z'/\langle r \rangle$ can be regarded as one equivalent scale to

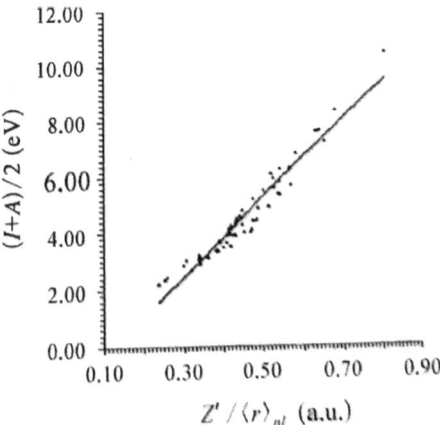

Fig. 4.3 The linear relationship between $(I + A)/2$ and $Z'/\langle r \rangle_{nl}$

Mulliken electronegativity. We call the ratio $Z'/\langle r \rangle$ the nuclear potential scale to electronegativity of the weakest bound electron.

Regarding the relation between $Z'/\langle r \rangle$ and Mulliken electronegativity, we did the following studies previously [185, 200]:

Mulliken electronegativity

$$X_M = (I + A)/2 \tag{4.5.8}$$

Absolute hardness [196]

$$\eta = (I - A)/2 \tag{4.5.9}$$

Gazquez-Ortiz relation [201]

$$\eta \approx \langle r^{-1} \rangle / 4 \tag{4.5.10}$$

Equations (4.5.8) and (4.5.9) together give

$$X_M \approx I - \eta \tag{4.5.11}$$

Then substituting it into Eq. (4.5.10), we get

$$X_M \approx I - \langle r^{-1} \rangle / 4 \tag{4.5.12}$$

We have investigated the relation between $Z'/\langle r \rangle$ and $I - \langle r^{-1} \rangle / 4$ for 72 elements in the periodic table, and a correlation coefficient equal to 0.986 was obtained (see Fig. 4.4 and Table 4.51).

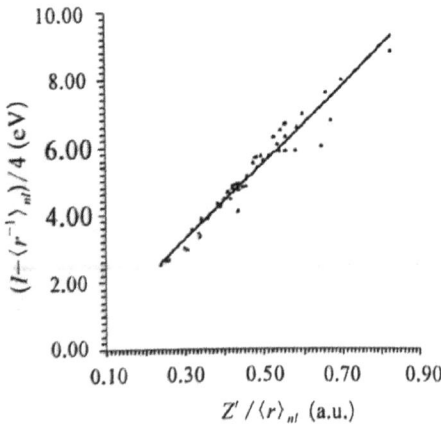

Fig. 4.4 The linear relation between $I - \langle r^{-1} \rangle_{nl}/4$ and $Z'/\langle r \rangle_{nl}$ for the top 72 elements in the periodic table (the lanthanides are excluded). (The energy unit of the vertical axis is converted to eV. The correlation coefficient is 0.986)

4.5 Electronegativity, Hard and Soft Acids and Bases …

Table 4.51 Values of $Z'/\langle r \rangle$

Z	Atom	Z'	<r> (a.u.)	Z'/<r> (a.u.)
1	H	1	1.5	0.666 7
2	He	1.071 9	0.964 4	1.111 3
3	Li	1.148 3	3.689 2	0.311 2
4	Be	1.312 3	2.517 6	0.521 2
5	B	1.008 9	2.292 8	0.440 0
6	C	1.057 1	1.825 7	0.579 0
7	N	1.116 3	1.527 7	0.730 6
8	O	0.901 5	1.399 5	0.644 1
9	F	1.007 7	1.228 0	0.820 5
10	Ne	1.105 1	1.093 6	1.010
11	Na	1.254 2	4.131 0	0.303 5
12	Mg	1.408 2	3.170 5	0.444 1
13	Al	1.160 9	3.390 0	0.342 4
14	Si	1.254 2	2.737 6	0.458 1
15	P	1.352 8	2.323 2	0.582 3
16	S	1.171 2	2.109 6	0.555 1
17	Cl	1.300 7	1.875 6	0.693 4
18	Ar	1.423 9	1.692 7	0.841 1
19	K	1.349 6	5.111 5	0.264 0
20	Ca	1.500 1	4.081 8	0.367 5
21	Sc	1.501 8	3.842 7	0.390 8
22	Ti	1.492 4	3.680 9	0.405 4
23	V	1.413 9	3.561 9	0.396 9
24	Cr	1.366 2	3.453 8	0.395 5
25	Mn	1.452 4	3.331 8	0.435 9
26	Fe	1.479 1	3.212 1	0.460 4
27	Co	1.428 0	3.127 4	0.456 5
28	Ni	1.344 2	3.060 6	0.439 1
29	Cu	1.320 7	2.987 2	0.442 1
30	Zn	1.563 7	2.864 5	0.545 8
31	Ga	1.185 8	3.439 9	0.344 7
32	Ge	1.304 1	2.900 5	0.449 6
33	As	1.414 9	2.549 4	0.554 9
34	Se	1.274 5	2.367 0	0.538 4
35	Br	1.408 8	2.157 5	0.652 9
36	Kr	1.538 7	1.987 0	0.774 3

(continued)

Table 4.51 (continued)

Z	Atom	Z'	<r> (a.u.)	Z'/<r> (a.u.)
37	Rb	1.410 9	5.494 2	0.256 8
38	Sr	1.559 1	4.494 4	0.346 9
39	Y	1.613 2	4.167 4	0.387 1
40	Zr	1.641 4	3.967 3	0.413 7
41	Nb	1.582 1	3.829 1	0.413 1
42	Mo	1.574 5	3.707 1	0.424 7
43	Tc	1.566 9	3.609 4	0.434 1
44	Ru	1.533 3	3.506 9	0.437 1
45	Rh	1.507 6	3.422 4	0.440 5
46	Pd	1.642 4	3.315 6	0.495 3
47	Ag	1.457 2	3.284 7	0.443 6
48	Cd	1.694 4	3.176 2	0.533 4
49	In	1.282 5	3.779 9	0.339 3
50	Sn	1.397 4	3.267 1	0.427 7
51	Sb	1.465 3	2.932 4	0.499 6
52	Te	1.404 8	2.734 1	0.513 8
53	I	1.506 4	2.529 9	0.595 4
54	Xe	1.630 7	2.356 9	0.691 8
55	Cs	1.490 5	6.137 9	0.242 8
56	Ba	1.640 9	5.087 6	0.322 5
57	La	1.730 0	4.997 7	0.346 1
72	Hf	1.698 4	3.995 3	0.425 0
73	Ta	1.843 9	3.833 4	0.480 9
74	W	1.806 0	3.729 3	0.484 2
75	Re	1.736 7	3.652 9	0.475 4
76	Os	1.858 9	3.529 8	0.526 6
77	Ir	1.894 5	3.441 4	0.550 4
78	Pt	1.830 7	3.379 9	0.541 6
79	Au	1.837 6	3.315 0	0.554 3
80	Hg	2.041 4	3.229 7	0.632 0
81	Tl	1.420 7	3.908 2	0.363 5
82	Pb	1.502 0	3.430 5	0.437 8
83	Bi	1.330 8	3.164 9	0.420 4
84	Po	1.427 0	2.939 3	0.485 4
85	At	1.467 8	2.752 9	0.533 1
86	Rn	1.596 3	2.581 4	0.618 3

Data source Zheng NW, Li GS. J Phys Chem, 1994, 98:3964

If the experimental data of ionization energy and electron affinity are used for fitting $(I - A)/2$ against $\langle r^{-1} \rangle$, a more accurate expression than Gazquez-Ortiz relation, i.e. Eq. (4.5.10) will be obtained.

Atoms have successive ionization energy and electron affinity, and the corresponding $\langle r^{-1} \rangle$ and $\langle r \rangle$ can be calculated theoretically, so the nuclear potential scale to the electronegativity of the weakest bound electron in the valence state can be completely calculated.

4.5.3 The Hard-Soft-Acid-Base Concept and Scale

In 1923, Lewis proposed an electron theory to describe acids and bases under the concept of electron pair donors and acceptors [202]. Based on this theory, any species that donates an electron pair is called a base, and any species that accepts an electron pair is called an acid. An acid can react with a base to form acid–base adduct by donating and accepting an electron pair. Compared with other definitions of acids and bases, this definition covers the most varieties of acids and bases, so it has the most extensive applications. The reactions of metal ions or atoms with organic or inorganic ligands to produce complexes or coordination polymers that we care about can be understood based on this theory.

In the complexes or coordination polymers, metal ions or atoms are electron pair acceptors, so they act as Lewis acids, while organic or inorganic ligands donate electrons from certain (or some) atom(s) or group(s) of atoms, so they act as Lewis bases. Those atoms or groups of atoms which directly donate electrons are called coordinating atoms or coordinating functional groups, and the elements, such as F, Cl, Br, I, O, S, Se, Te, N, P, As, etc., at the upper right corner of the extended periodic table often appear as coordinating atoms. Naturally there is a problem of strength of the electron donors and acceptors relationship between metal ions or atoms and coordinating atoms or functional groups, and such strength will be reflected by the stability of complexes or coordination polymers. Some metal ions (or atoms) form very stable complexes with some coordinating atoms, but with some other coordinating atoms the formed complexes have very low stability, however, it is totally opposite for some other metal ions (or atoms). Take, for example, halide complexes of Fe^{3+} and Hg^{2+} (See Table 4.52).

Table 4.52 The stability constants of halide complexes of Fe^{3+} and Hg^{2+} (lgk_1) central ions, ligands

Central ion	Ligand			
	F$^-$	Cl$^-$	Br$^-$	I$^-$
Fe^{3+}	6.04	1 41	0.49	–
Hg^{2+}	1.03	6 72	8.94	12.87

Data source Dai AB, et al. Coordination chemistry. Science Press, Beijing, 1987

From Table 4.52 we can see that the stability of the complexes formed by the central metal ion Fe^{3+} with halogens as coordinating atoms changes in the following way with the type of halogen atoms, i.e. F >> Cl > Br > I, however, the stability of the complexes formed by them and central metal ion Hg^{2+} is exactly opposite, i.e. F << Cl < Br < I. Similar trend also appears when coordinating atoms come from other groups, i.e. O >> S > Se > Te and O << S ~ Se ~ Te; N >> P > As > Sb > Bi and N << P > As > Sb > Bi. Thus, Abrland et al. divided the central metal ions into two classes (a) and (b) [203–205]. The stability of the complexes formed by (a) class central atoms and ligands has the following relation with the change of coordinating atoms: N >> P > As > Sb > Bi, O >> S > Se > Te and F >> Cl > Br > I; while the relation for (b) class is N << P > As.Sb > Bi, O << S ~ Se ~ Te and F << Cl < Br < I. The central atoms don't all behave as classically as Fe^{3+} and Hg^{2+}, so the classification into (a) class and (b) class is qualitative and relatively crude. Later Pearson further investigated the raw of the stability of complexes, and proposed the concept and classification of hard and soft acids and bases [203, 206]. The old (a) class central atoms fall into the category of hard acids, the (b) class central atoms belong to the category of soft acids, and some non-classical Lewis acids become members of the borderline acids which are at the middle of hard acids and soft acids. For Lewis bases which donate electron pairs, the corresponding classification has also been given: soft bases, borderline bases, and hard bases. Tables 4.53 and 4.54, respectively, list the classifications for some common acids and bases. Later based on plenty of experimental information, such as equilibrium constant, bond energy, rate constant, etc., Hard and Soft Acids and Bases (HSAB) theory was summarized. In this theory, it is believed that hard acids tend to form complexes with hard bases and soft acids tend to form complexes with soft bases [203, 207]. HSAB theory is an empirical theory, but it has very wide applications [208].

Many researchers have focused on the quantitative studies of the strength (or scale) of HSAB [202, 203, 209–213]. They are not perfect, but there is no doubt that

Table 4.53 Some common hard acids, borderline acids and soft acids

Hard acid
H^+, Li^+, Na^+, K^+, R_b^+, Cs^+, Be^{2+}, $Be(CH_3)_2$, Mg^{2+}, Ca^{2+}, Sr^{2+}, Ba^{2+}, rare earth metal ions RE^{n+} ($n = 2, 3, 4$; RE = Sc, Y, La, Ce, Pr, Nd, Pm, Sm, Eu, Gd, Tb, Dy, Ho, Er, Tm, Yb, Lu), UO_2^{2+}, U^{4+}, Th^{4+}, Pu^{4+}, Ti^{4+}, Zr^{4+}, Hf^{4+}, Cr^{3+}, Cr^{6+}, MoO^{3+}, WO^{4+}, Mn^{2+}, Mn^{7+}, Fe^{3+}, C_O^{3+}, BF_3, BCl_3, $B(OR)_3$, Al^{3+}, $AlCH_3$, AlH_3, $Al(CH_3)_3$, RCO^+, CO_2, NC^+, Si^{4+}, Sn^{4+}, CH_3Sn^{3+}, $(CH_3)_2Sn^{2+}$, N^{3+}, RPO_2^+, $ROPO_2^+$, $As^{3+}RSO_2^+$, $ROSO_2^+$, SO_3, I^{5+}, I^{7+}, Cl^{3+}, Cl^{7+}, HX (hydrogen bonding molecules, 含 H_2O)
Soft acid
Cu^+, Ag^+, Au^+, Cd^{2+}, Hg^+, Hg^{2+}, CH_3Hg^+, Pd^{2+}, Pt^{2+}, Pt^{4+}, Tl^{3+}, Tl^+, $Tl(CH_3)_3$, $Ga(CH_3)_3$, $GaCl_3$, $GaBr_3$, GaI_3, BH_3, $Co(CN)_2^{3-}$, HO^+, RO^+, RS^+, RSe^+, Te^{4+}, RTe^+, Br_2, Br^+, I_2, I^+, ICN, O, Cl, Br, I, N, RO •, RO_2 •, $M°$ (metal atoms), trinitrobenzene, tetrachlorobenzoquinone, quinines 等, tetracyanoethylene 等, carbenes

(continued)

Table 4.53 (continued)

Border acid
Fe^{2+}, Co^{2+}, Ni^{2+}, Cu^{2+}, Zn^{2+}, Pb^{2+}, Sn^{2+}, Sb^{3+}, Bi^{3+}, Rh^{3+}, Ir^{3+}, Ru^{2+}, Os^{2+}, $B(CH_3)_3$, SO_2, NO^+, R_3C^+, $C_6H_5^+$, GaH_3

Data source
(1) Huheey JE. Inorganic chemistry: principles of structure and reactivity, 3rd edn. Harper International sl Edition, Cambridge, 1983: 314
(2) Zhang XL, Kang H. Coordination chemistry. Central South University of Technology Press, Changsha, 1986: 323
(3) Dai AB, et al. Coordination chemistry. Science Press, Beijing, 1987: 280
(4) Huang CH. Rare earth coordination chemistry. Science Press, Beijing, 1997: 13
Remark In the paper: Xu GX, Gong S, Huang CH, et al. Prog Nat Sci, 1993, 3:1, the following order of increasing softness of metal elements was given: rare earth elements (the hardest acids) < pre-transition elements (such as Ti, Zr, Hf, V, Nb, ta, Cr, Mo, W) < mid-transition elements (such as Mn, Tc, Re, Fe, Ru, Os, Co, Rh, Ir, Ni, Pd, Pt) < post-transition elements (such as Cu, Ag, Au, Zn, Cd, Hg) and metals that follow transition metals (such as Tl, In, Pb, Sn, Bi)

Table 4.54 Some common hard bases, borderline bases and soft bases

Hard base
F^-, Cl^-, O^{2-}, OH^-, H_2O, CH_3COO^-, ClO_4^-, NO_3^-, SO_4^{2-}, PO_4^{3-}, CO_3^{2-}, RO^-, ROH, R_2O, NH_3, RNH_2, N_2H_4
Soft base
I^-, S^{2-}, RS^-, RSH, R_2S, $S_2O_3^{2-}$, SCN^-, R_3P, R_3As, $(RO)_3P$, CN^-, RNC, CO, C_2H_4, C_6H_6, H^-, R^-
Border base
$C_6H_5NH_2$, C_5H_5N, N_3^-, Br^-, NO_2^-, SO_3^{2-}, N_2

Data source
(1) Huheey JE. Inorganic chemistry: principles of structure and reactivity, 3rd edn. Harper International sl Edition, Cambridge, 1983: 314
(2) Zhang XL, Kang H. Coordination chemistry. Central South University of Technology Press, Changsha, 1986: 323
(3) Dai AB, et al. Coordination chemistry. Science Press, Beijing, 1987: 280
Remark (a) In the literature: Maksic ZB. The concept of the chemical bond. Spinger-Verline, Berlin, 1990: 65, the following order of hardness was given: $F^- > Cl^- > Br^- > I^-$; $OH^- > SH^- > SeH^- >$; $NH_2^- > PH_2^-$; $CH_3^- > SiH_3^-$; $F^- > OH^- > NH_2^- > CH_3^-$
(b) In the paper: Xu GX, Gong S, Huang CH, et al. Prog Nat Sci, 1993, 3:1, the order of hardness of bases was given as follows: Oxygen ligand is hard base, S ligand is soft base, and other ligands such as N, Cl, P, etc. are between these two bases

electronegativity is correlated with the hardness of Lewis acids for central ions (or atoms) [185, 203, 210, 213]. In this regard, there are formulae from literatures

$$\begin{cases} X_M = \frac{I+A^{[201]}}{2} \\ \eta = \frac{I-A}{2} \end{cases} \quad (4.5.13)$$

or

$$X_M = I - \eta \qquad (4.5.14)$$

In the above two equations, X_M is Mulliken electronegativity, I and A are ionization energy and electron affinity, respectively, and η is the absolute hardness of Lewis acids [189, 215].

$$\mathscr{Z} = \frac{Z}{r_k^2} - 7.7X_z + 8.0 \qquad (4.5.15)$$

In this equation which has two parameters, $\mathscr{Z}$ represents the scale to the strength of a Lewis acid, parameter Z/r_k^2 is related with the electrostatic force (Z is the nuclear charge of the atomic core and r_k is the radius of an ion), and another parameter X_Z, which is correlated with the strength of a covalent bond, is the electronegativity of elements in valence states proposed by Zhang. The hardness calculated by these equations has a certain reference value. However, for the discussions later in this book, the calculated values for hardness of trivalent lanthanide ions using Eqs. (4.5.13) and (4.5.15) are all relatively low. Actually they are quite hard acids.

4.5.4 Molecular Design of Coordination Polymers

In 1967, for the first time Pedersen successfully synthesized the crown ethers, and investigated their selective binding behavior with metal ions, thus opened a new field for the study of macrocyclic compounds [214]. Crown ethers can selectively bind metal ions very like the reaction between enzymes and substrates in biology, indicating that it has the function of molecular recognition. Cram synthesized and studied a series of optically active crown ethers, and used crown ethers as hosts (acceptors) to selectively bind ions or molecules which are treated as guests (substrates), therefore, established Host–Guest Chemistry [215]. Macrocyclic compounds not only can recognize metal ions or molecules, but also can be assembled by non-covalent binding forces between molecules to form aggregates with new properties and new functions. The assembled aggregates are not single molecules anymore and the driving forces of the assembly are non-covalent forces which is different from the covalent bonding inside the molecules. The study of molecular recognition and assembly process is beyond the molecular level. Lehn proposed a concept of Supramolecular Chemistry during the study of molecular recognition and assemblies [216]. They three were awarded the 1987 Nobel Prize in Chemistry for their pioneering contributions to this field.

The so-called molecular recognition is the process that a host (acceptor) selectively binds a guest (substrate) to form a specific structure and function. The so-called Supramolecular Chemistry is the chemistry that more than two molecules assemble beyond the molecular level by non-covalent forces between them. The study of

4.5 Electronegativity, Hard and Soft Acids and Bases ...

molecular recognition and assemblies started from solution system, later extended to solids. The term crystal engineering was first proposed by Schmidt [217]. This term soon was associated with Supramolecular Chemistry, and formed two fields, organic molecular networks and coordination polymers, respectively, so the research on organic molecular networks and coordination polymers can be regarded as the new development of Supramolecular Chemistry.

The so-called coordination polymer is a polymer of three-dimensional crystal structure assembled by non-covalent interactions between modular building units which expand repeatedly in the space and are formed by coordination bonds between metal ions and organic or inorganic ligands. Because it is assembled crystal, it is called crystal engineering, and also because it is assembled beyond the molecular level, it is regarded as the new development of Supramolecular Chemistry.

The composition and structure properties of a coordination polymer determine that it has obvious advantages: first, the metal ion is its first component and basic structural element. In the coordination and assembly environment, the metal ion may give its properties, such as properties of electricity, magnetism and optics, the acid–base properties of chemistry, oxidation–reduction properties, catalysis properties, coordination properties, etc. to the designed and synthesized coordination polymers, such that it becomes versatile and varied; second, the second component and basic structural element is organic or inorganic ligand. For a wide variety of ligands, especially organic ligands, their compositions, spatial structures, softness and hardness of coordinating atoms, various ways of coordinating with metal ions, etc. are inexhaustible, and the constructed modules and spatial expansions of modules are limitless; third, under different conditions the assembled structures, properties and functions are different. These evident advantages of coordination polymers provide unlimited imagination space for the design and synthesis of coordination polymer materials. They also make the size, shapes, types, surface area, acid–base properties of inner surface, hydrophobic and hydrophilic properties, catalysis properties (acid–base catalysis, redox catalysis, chiral catalysis, and absorption and desorption of the hole), etc. of the pores in coordination polymer materials highly adjustable, which is hard to be matched by zeolitic materials, inorganic porous zeolitic materials, and inorganic and organic crystalline materials. At the same time, also because of these evident advantages, coordination polymer materials have wide application potential in the fields of catalysts, molecular sieves, molecular magnets, nonlinear optical elements, sensors, drug synthesis, etc. Therefore, the study of coordination polymers has received extensive attentions from scientists in different countries. People use metal ions and organic or inorganic ligands to design and synthesize massive coordination polymer crystals of different types such as zigzag structure, brick wall structure, ladder-like structure, helical structure, honeycomb structure, diamond structure, etc. [218–255].

The molecular design of coordination polymers is a very big research topic. It has been discussed in many reviews and books, for instance, Xu et al. [256] gave a feature review on molecular design, synthesis, structure and bonding of polynuclear rare-earth complexes and clusters, and Huang [257] published a book *Rare Earth Coordination Chemistry*, which are worth reading carefully. Here, we introduce some of our thoughts and practices as follows only based on the work in our lab:

(1) HSAB matching. It was pointed out in Ref. [256] that HSAB principle is one rule for the molecular design of complexes and coordination polymers. At the same time, this reference gave the order of increasing softness of metal elements: rare earth elements (the hardest acids) < pre-transition elements < mid-transition elements < post-transition elements and metals that follow transition metals (see the notes in Table 4.53), as well as the conclusion that oxygen atoms tends to bind rare earth ions and nitrogen atoms tends to bind transition metal ions. Reference [257] more vividly and more specifically reinforced the above conclusion using the classification data of chemical bonds of rare earth elements which have structural data published from 1935 to the first of 1995. Among 719 types of complexes, 587 types belong to RE-O and only 27 types belong to RE-N. So oxygen atoms tend to bind rare earth ions. We note that among all coordination polymers which have been synthesized until 2000, most of the metal ions are post-transition elements, such as Cu, Ag, Au, Zn, Cd and Hg, and mid-transition elements, such as Pd, Pt, Ru, Fe, Co, Ni, all of which are relatively soft Lewis acids, and they match non-oxygen ligands which are relatively soft Lewis bases (most of them are nitrogen-containing ligands), however, coordination polymers formed by rare earth metal ions which are hard Lewis acids and oxygen-containing ligands which are hard Lewis bases have seldom been investigated. So we started the synthesis of the rare-earth metal coordination polymers with oxygen-containing ligands relatively early [228, 229, 258].

(2) The concept and structure of pendent functional groups or functional atoms in the framework structure. When choosing compounds which contain coordinating atoms O and N as ligands and rare earth ions as central elements, there may be two types of binding conditions. The first condition is that rare earth ions likes to bind oxygen and relatively don't like nitrogen, but when O and N atoms in the ligands form a ring structure with rare earth ions, it is possible for N atom to bind rare earth ions simultaneously with O atom due to the stability brought by forming a ring (see Fig. 4.5). In the figure, the coordination polymer is $[Ce_2(Hpdc)_3(H_2O)_4] \cdot 2H_2O$. $Hpdc^{2-}$ represents 3,5-pyrazoledicarbonxylic acid anion. The N atom of the pyrazole and a neighboring carboxyl oxygen atom form a five-membered ring with Ce^{3+}, thus, N and O atoms are involved in the binding simultaneously. Specifically for Ce1, three $Hpdc^{2-}$ ligands individually bind Ce1 to form three five-membered rings through a carboxyl oxygen atom and its neighboring nitrogen atom on a ring, i.e. O3/N2, O5/N3 and O9/N5 labeled in the figure; three other $Hpdc^{2-}$ ligands bind Ce1 individually through a carboxyl oxygen, i.e. O2, O7 and O12 in the figure. The coordination number of Ce1 is 9. For Ce2, one carboxyl from one $Hpdc^{2-}$, i.e. O1 and O2 in the figure,

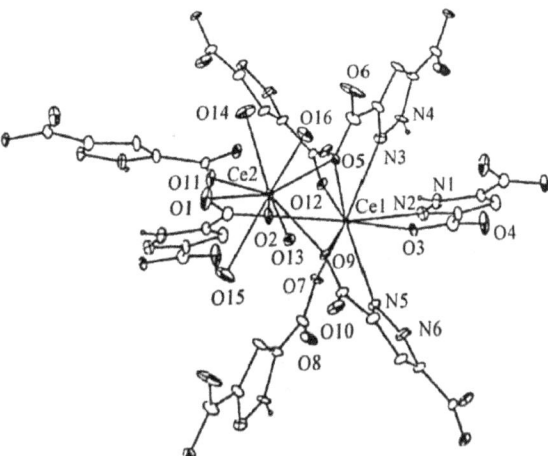

Fig. 4.5 Coordination in [Ce$_2$(Hpdc)$_3$(H$_2$O)$_4$]·2H$_2$O (Reproduced from Pan L, Huang XY, Li J et al (2000) Angew Chem Int Ed 39:527)

form a ring by binding Ce2, three other Hpdc^{2-} individually bind Ce2 through a carboxyl oxygen, i.e. O5, O9 and O11, and the remaining four coordinating positions are taken by oxygen atoms of four H$_2$O. The coordination number of Ce2 is also 9. In the second condition, rare earth ions only bind O but not N. N atoms or N-containing functional groups can't enter the framework structure instead they are dangling from the frame structure (see Figs. 4.6 and 4.7). In both figures, rare earth ions only bind oxygen but not nitrogen. The non-coordinating N-containing –NH$_2$ groups are dangling from the framework structure. From the host–guest chemistry standpoint, the dangling N atoms or N-containing functional groups are atoms or functional groups with functions. Therefore, we proposed to design dangling functional groups or functional atoms in the framework structure, which is helpful for the realization of functions for host–guest chemical reactions.

(3) Use the method of ligand modifications to convert rare earth coordination polymers from a close-packed structure to an open porous structure. The combination of hard acids with hard bases is a strong ionic chemical bond pattern, and the combination of soft acids and soft bases is a strong covalent chemical bond pattern. Covalent bonds are directional, but ionic bonds are non-directional, thus, ionic crystal structures are close-packed structures, and rare-earth metal coordination polymers with oxygen-containing ligands can easily have close-packed structures. We proposed to introduce some groups to modify ligands so that it may have "open" effect, in order to convert a close-packed structure into a porous structure (see Fig. 4.8) [236, 259, 260].

(4) π-bond structures. For dicarboxylic acid ligands which contain benzene, pyridine, pyrazole, or pyrazine, there are conjugated π-bonds between the ring and carboxyl group CO_2^-, so when carboxyl oxygen binds the central ion, the

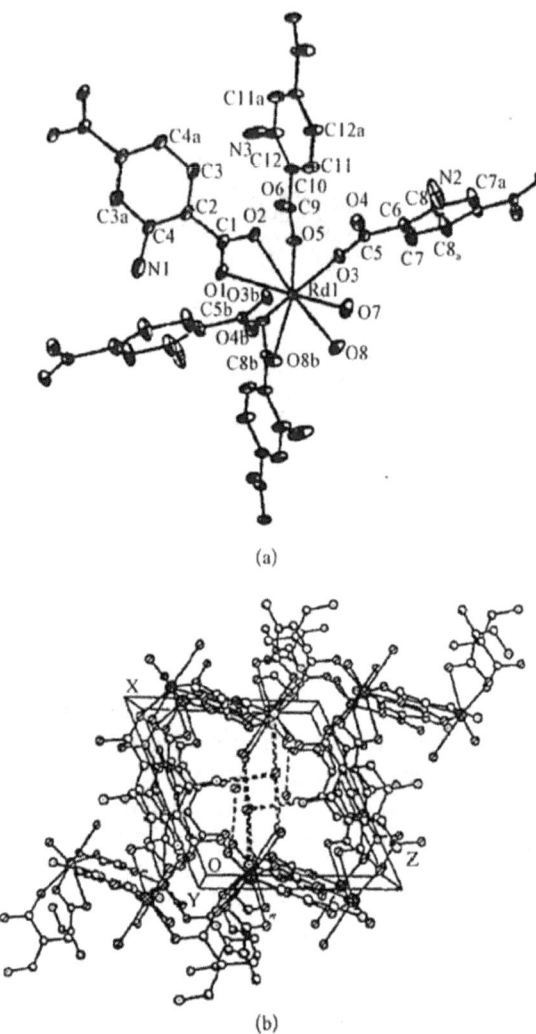

Fig. 4.6 [Nd(C$_8$H$_5$NO$_4$)$_{1.5}$(H$_2$O)$_2$]·2H$_2$O (Reproduced from Xu HT, Zheng NW, Jin XL et al (2003) J Mol Struct 654:183). C$_8$H$_5$NO$_4^{2-}$ represents 2-aminoterephthalic acid anion

chelation of central ion with the ring and CO_2^- or the bridging plane must be maintained at a certain range of angle, otherwise the conjugated π structure will be destroyed [235, 260].

(5) The coordination number of rare-earth metal coordination polymers with oxygen-containing ligands. It was pointed out in Ref. [256] that the most common coordination number for rare-earth metal ion is 8, and the next is 9. In Ref. [257] it was also pointed out that rare earth ions have large ionic radius, so the bonds formed during the complex formation are not highly directional, and

4.5 Electronegativity, Hard and Soft Acids and Bases ...

Fig. 4.7 Coordination polymer [La(C$_8$H$_5$NO$_4$)$_{1.5}$(H$_2$O)]$_n$ (Reproduced from Xu HT, Zheng NW, Jin XL et al (2002) Chem Lett 1144). C$_8$H$_5$NO$_4^{2-}$ represents 5-aminoisophthalic acid anion

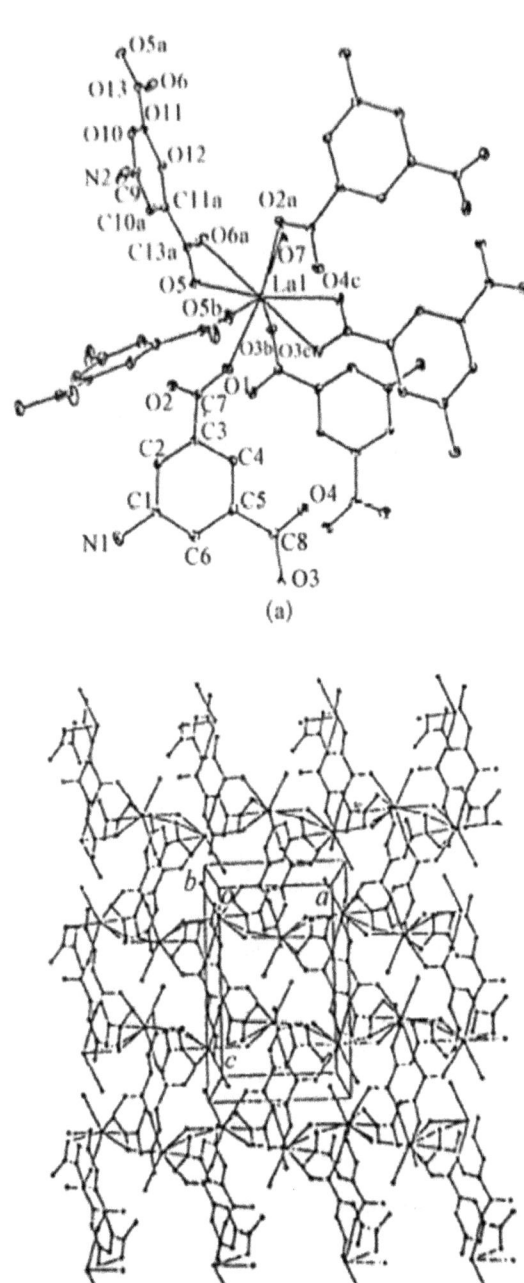

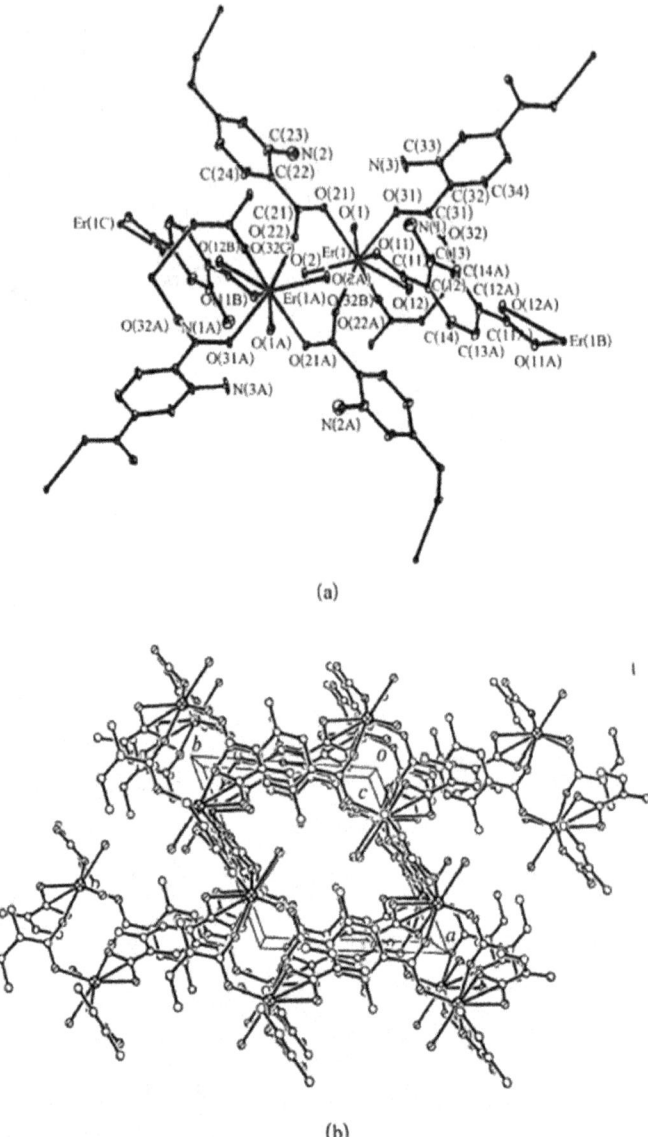

Fig. 4.8 Open framework structure of $Er_2(abdc)_3 \cdot 5.5H_2O$ (Reproduced from Wu YG, Zheng NW et al (2002) J Mol Struct 610:181). $abdc^{2-}$ represents 2-amino-1,4-benzenedicarboxylic acid anion

the coordination number can vary in the range of 3–12. Among them the most are the coordination polymers with coordination number equal to 8. Indeed, the coordination number of rare earth ions in the rare-earth metal complexes and coordination polymers are affected by many factors, and the variation range can be large. The coordination number is very high in the lanthanide nitrate complexes. For instance, in $(NH_4)_2La(NO_3)_5 \cdot 4H_2O$, the coordination number of La^{3+} is 12 [257]. If $C_2O_4^{2-}$ is used to replace NO_3^-, the coordination number becomes smaller. In $Nd_2(C_2O_4)_3 \cdot 10H_2O$, the coordination number of Nd^{3+} is 9 [257]. Why? The charge that $C_2O_4^{2-}$ carries is more negative than that of NO_3^-, which causes the repulsive forces between ligands increase, so the coordination number decreases. Besides ligand charge, other factors such as steric hindrance (including the size of the ligand, ligand configuration, ligand rigidity and flexibility), ring effect, the type of assembly between modules, etc. all affect the coordination number of the central ion in complexes and coordination polymers. In general, most rare-earth metal coordination polymers with oxygen-containing ligands have coordination number equal to 8, and the next is 9. When neutral small molecules, such as H_2O, are involved in the binding, the coordination number may get bigger, for instance, up to 10. Contrary to rare-earth metal coordination polymers, the coordination number of the mid-transition metal may get bigger in its coordination polymer. For example, Mn^{2+} has coordination number equal to 6 in complexes, however, in $[Mn(3,5\text{-pdc}) \cdot 2H_2O]$, the coordination number of Mn^{2+} is 7 which is rare (see Fig. 4.9) [261].

(6) The idea of nanoparticle preparation by thermolysis of coordination polymers as precursors. Coordination polymers consist of metal ions and organic or inorganic ligands and will be all pyrolyzed after being heated to a certain temperature. We note that when coordination polymers were pyrolyzed as precursors, nanoparticles with well distributed grain size were obtained [236].

(7) When ligands which contain both O and N coordinating atoms bind the mixture of rare earth metal ions and transition metal ions, rare earth ions act different from the first condition described in (2) previously. Transition metal ions prefer to bind N, while rare earth ions are harder Lewis acids and prefer to bind O which is harder base. If there is a condition to form a ring, transition metal ions have preference to form a ring with O and N, while rare earth ions only bind O (including O from ligands as well as O from solvent H_2O), and it is very hard to synthesize if rare earth ions bind O and N simultaneously.

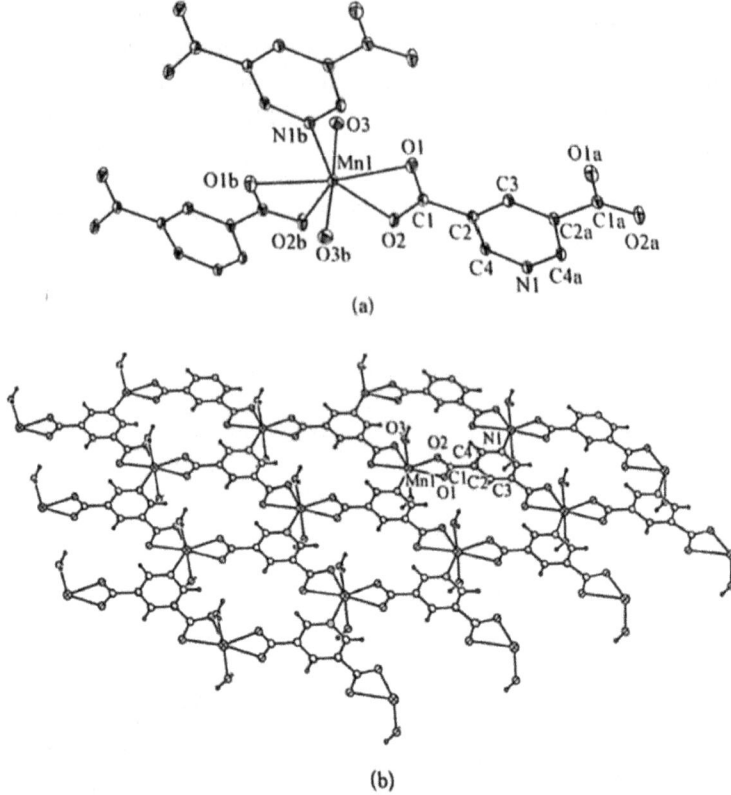

Fig. 4.9 The coordination number of Mn^{2+} is 7 in coordination polymer [Mn(3,5-pdc)·2H$_2$O] (Reproduced from Xu HT, Zheng NW, Xu HH et al (2002) J Mol Struct 610:47). 3,5-pdc^{2-} represents 3,5-pyridine dicarboxylic acid anion

References

1. Zheng NW (1988) A new introduction to atoms. Nanjing Education Press, Nanjing
2. Zheng NW (1986) KEXUE TONGBAO 31:1238–1242; Zheng NW (1985) Chin Sci Bull 30:1801–1804
3. Zheng NW (1987) KEXUE TONGBAO 32:1263–1267; Zheng NW (1986) Chin Sci Bull 31:1316–1319
4. Zheng NW (1988) KEXUE TONGBAO 33:916–920; Zheng NW (1987) Chin Sci Bull 32:354–357
5. Zheng NW (1977) Chin Sci Bull 22:531–535
6. Zheng NW, Zhou T, Wang T et al (2002) Phys Rev A 65:052510
7. Zheng NW, Wang T (2003) Chem Phys Lett 376:557; (2003) Int J Quantum Chem 93:334; (2004) Int J Quantum Chem 98:495
8. Wang T (2003) The study of WBEPM Theory with respect to transition, ionization and energy level of many-electron atoms and ions. University of Science and Technology of China, Hefei
9. Zhou T (2001) The study of WBEPM Theory with respect to atomic energy level and ionization energy. University of Science and Technology of China, Hefei

10. Zheng NW, Xin HW (1991) J Phys B: At Mol Opt Phys 24:1187.
11. Thewlis T (1961) Encyclopaedic dictionary of physics, vol 2. Pergamon Press, Oxford. In this reference, the ionization energy was defined as follows: For a free particle like an atom or a molecule, the energy required to completely remove the weakest bound electron from the particle in its ground state such that the resulting (positive) ion is also in its ground state, is called the ionization energy. Thus, the energies required to successively ionize a neutral particle is called the first, the second, the third, ..., ionization energy of the particle
12. Cowan RD (1981) The theory of atomic structure and spectra. University of California Press, Berkeley. In this reference, the energy difference between a ground-state atom (or ion) and the first ionization limit (i.e. the energy difference between a ground-state ion and a higher order ground-state ion in the successive ionization) is called the ionization energy of the atom (or the ion). At the same time, the relation between the total electronic energy E of the ground-state ion at the mth ionization stage with successive ionization energy I_j was given as follows:

$$E = -\sum_{j=m}^{Z} I_j (1 \leq m \leq Z)$$

13. Zheng LL, Xu GW (1988) Atomic structure and atomic spectra. Peiking University Press, Beijing, p 123. In this reference, two ways were given to represent the ionization energy: one is by subtracting the energy of the lowest energy level of the atom from the energy of the lowest energy level of the ion, which gives rise to the ionization energy, and this is also a common method to calculate ionization energy; Another is by subtracting the average energy of the ground-state electronic configuration of the atom from the average energy of the ground-state electronic configuration of the ion, and the resulting energy is the ionization energy
14. Xu KZ (2000) Advanced atomic and molecular physics. Science Press, Beijing, p 158. In this reference, the adiabatic ionization energy is defined as the energy difference of the minimum on the potential curve between a molecule and its ion after one-time ionization. It is different from the ionization energy (vertical ionization energy) in Koopmans approximation
15. Edlen B (1964) Encyclopedia of physics, vol 27. Springer-Verlag, Berlin
16. Parpia FA, Fischer CF, Grant IP (1996) Comput Phys Commun 94:249
17. Kim YK, Martin WC, Weiss AW (1988) J Opt Soc Am B 5:2215
18. Chen MH, Cheng KT, Johnson W (1993) Phys Rev A 47:3692
19. Dzuba VA, Flambaum VV, Sushkov OP (1983) Phys Lett A 95:230; (1995) Phys Rev A 51:3454
20. Safronova MS, Johnson WR, Safronova UI (1996) Phys Rev A 53:4036
21. Dzuba VA, Johnson WR (1998) Phys Rev A 57:2459
22. Eliav E, Kaldor U, Ishikawa Y (1996) Phys Rev A 53:3050
23. Jönsson P, Fischer CF, Godefroid MR (1999) J Phys B: At Mol Opt Phys 32:1233
24. Safronvoa UI, Johnson WR, Safronova MS et al (1999) Phys Scr 59:286
25. Eliav E, Vilkas MJ, Ishikawa Y et al (2005) Chem Phys 311:163
26. Dzuba VA (2005) Phys Rev A 71:032512
27. Safronvoa UI, Johnson WR, Safronova MS (2007) Phys Rev A 76:042504
28. Safronvoa UI, Johnson WR, Berry HG (2000) Phys Rev A 61:052503
29. Fischer CF, Tachiev G, Gaigalas G et al (2007) Comput Phys Commun 176:559
30. Jursic BS (1997) Int J Quantum Chem 64:255
31. Davidson ER, Hagstron SA, Chakravorty SJ et al (1991) Phys Rev A 44:7071; (1993) 47:3649
32. Edlen B (1971) Topics in modern physics: a tribute to Edward U. Condon. Colorado Associated University Press, Colorado
33. Faktor MM, Hanks R (1969) J Inorg Nucl Chem 31:1649
34. Sugar J, Reader J (1973) J Chem Phys 59:2083
35. Sugar J (1975) J Opt Soc Am 65:1366
36. Sinha SP (1975) Helv Chim Acta 58:1978
37. Sinha SP (1977) Inorg Chim Acta 22:L5
38. Sinha SP (1978) Inorg Chim Acta 27:253
39. Zheng NW, Zhou T, Yang RY et al (2000) Analysis of the bound odd-parity spectrum of krypton by weakest bound electron potential model theory. Chem Phys 258:37–46

40. Zheng NW, Ma DX, Yang RY et al (2000) An efficient calculation of the energy levels of the carbon group. J Chem Phys 113:1681
41. Zheng NW, Wang T, Ma DX et al (2002) Simple method for the precise calculation of atomic energy levels of IB elements in the periodic table. Int J Quantum Chem 87:293
42. Ma DX, Zheng NW, Lin X (2003) Study on energy levels of atom neon. Spectrochimca Acta: Part B 58:1625–1645
43. Zheng NW, Li ZQ, Fan J et al (2002) Calculation of high Rydberg levels of atom Zn with the WBEPM theory. J Phys Soc JPN 71:2677–2680
44. Zheng NW, Li ZQ, Ma DX et al (2004) Can J Phys 82
45. Fan J, Zheng NW, Ma DX et al (2004) Calculation of the energy levels to high states in atomic oxygen. Phys Scr 69:398
46. Zheng NW, Fan J, Ma DX et al (2003) Theoretical study of energy levels and transition probabilities of singly ionized aluminum (Al II). Phys Soc JPN 72:3091–3096
47. Ma DX, Zheng NW, Fan J (2004) Theoretical analysis on 3dnl J=1e–5e autoionizing levels in Ca. J Phys Chem Ref Data 33:1013
48. Zhang TY, Zheng NW, Ma DX (2007) Theoretical calculation of energy levels of Sr I. Phys Scr 75:763
49. Zhou T (2001) Study on atomic energy level and ionization energy within WBEPM Theory. University of Science and Technology of China, Hefei
50. Ma DX (2005) The development of WBEPM Theory and the study of its application. University of Science and Technology of China, Hefei
51. Blundell SA, Johnson WR, Safronova MS et al (2008) Relativistic many-body calculations of the energies of n = 4 states along the zinc isoelectronic sequence. Phys Rev A 77:032507
52. Safronova UI, Cowan TE, Safronova MS (2006) Excitation energies, hyperfine constants, E1 transition rates and lifetimes of $4s^2nl$ states in neutral gallium. J Phys B: At Mol Opt Phys 39:749
53. Safronova UI, Safronova MS, Johnson WR (2005) Excitation energies, hyperfine constants, E1, E2, and M1 transition rates, and lifetimes of $6s^2nl$ states in Tl_I and Pb_{II}. Phys Rev A 71:052506
54. Safronova UI, Safronova MS (2004) Can J Phys 82:743
55. Safronova UI, Johnson WR (2004) Excitation energies, oscillator strengths, and lifetimes of levels along the gold isoelectronic sequence. Phys Rev A 69:052511
56. Safronova UI, Johnson WR, Safronova MS et al (2002) Relativistic many-body calculations of energies for core-excited states in sodiumlike ions. Phys Rev A 66:042506
57. Safronova UI (2000) Excitation energies and transition rates in Be-, Mg-, and Zn-like ions. Mol Phys 98:1213–1225
58. Bieron J, Fischer CF, Godefroid M (2002) Hyperfine-structure calculations of excited levels in neutral scandium. J Phys B: At Mol Opt Phys 35:3337
59. Berengut JC, Dzuba VA, Flambaum VV et al (2004) Configuration-interaction calculation for the isotope shift in Mg I. Phys Rev A 69:044102
60. Dzuba VA (2005) Phys Rev A 71:062501; (2005) 71:032512
61. Dzuba VA, Sushkov OP, Johnson WR et al (2002) Energy levels and lifetimes of Gd IV and enhancement of the electron electric dipole moment. Phys Rev A 66:032105
62. Safronova MS, Johnson WR, Safronova UI (1997) J Phys B: At Mol Opt Phys 30:2375; (1996) Phys Rev A 54:2850
63. Parr RG, Yang WT (1989) Density-functional theory of atoms and molecules. Oxford University Press, New York
64. Nagy A (1998) Phys Rep-Rev Sec Phys Lett 298:2
65. Kozlov MG, Porsev SG, Flambaum VV (1996) Manifestation of the nuclear anapole moment in the M1 transitions in bismuth. J Phys B: At Mol Opt Phys 29:689–697
66. Lauderdale WJ, Stanton JF, Gauss J et al (1992) Restricted open-shell Hartree-Fock-based many-body perturbation theory: theory and application of energy and gradient calculations. J Chem Phys 97:6606
67. Seaton MJ (1966) Quantum defect theory I. General formulation. Prog Phys Soc 88:801–814

68. Lu KT (1971) Spectroscopy and collision theory. The Xe absorption spectrum. Phys Rev A 4:579
69. Fano U (1975) Unified treatment of perturbed series, continuous spectra and collisions. J Opt Soc Am 65:979–987
70. Fischer CF (1990) Variational predictions of transition energies and electron affinities: He and Li ground states and Li, Be, and Mg core-excited states. Phys Rev A 41:3481
71. Dzuba VA, Flambaum VV, Kozlov MG (1996) Combination of the many-body perturbation theory with the configuration-interaction method. Phys Rev A 54:3948
72. Zheng LM, Xu GW (1988) Atomic structure and atomic spectroscopy. Peiking University Press, Beijing
73. Xu KZ (2000) Advanced atomic and molecular physics. Science Press, Beijing
74. Cowan RD (1981) The theory of atomic structure and spectra. University of California Press, Berkeley
75. Slater J C (1960) Quantum theory of atomic structure, vol 1. McGraw-Hill Book Company Inc., New York, pp 17–19. Ritz Combination Principle indicates that in any atomic spectra, the frequency of an observed spectral line can be written as the difference of two spectral terms in unit of frequency
76. Langer RM (1930) A generalization of the Rydberg formula. Phys Rev 35:649
77. Connerade JP (2003) Highly excited atoms (trans: Zhan MS, Wang J). Science Press, Beijing, pp 27–28. In this book it is pointed out that for many-electron atoms or molecules, or even any near spherical dense collection of charged particles, Rydberg series is still shown in energy levels even when inside excitation exists, i.e. the infinite series of the energy level E_n clearly obey the formula as follows $E_{nl} = E_\infty - \frac{R_M Z^2}{(n-\mu_l)^2} = E_\infty - \frac{R_M Z^2}{n^{*2}}$
78. Martin WC (1980) Series formulas for the spectrum of atomic sodium (Na i). J Opt Soc Am 70:784–788
79. Zhang GY, Xue LP, Xia T et al (2007) J At Mol Phys 24:1104. (The maximum absolute deviation is 1.91 cm^{-1} and the maximum relative deviation is 4×10^{-5})
80. Xue LP, Zhang GY, Zhang XL et al (2005) J At Mol Phys 22:747. (Most deviations $\leq$ 1 cm^{-1} and the maximum deviation is -18 cm^{-1})
81. Xue LP, Zhang GY, Yin Z et al (2006) J At Mol Phys 23:1133. (Most deviation $\leq$ 1 cm^{-1}, the maximum relative deviation is 0.032% and the absolute deviation is 484.7 cm^{-1})
82. Zhang GY, Xue LP, Zhang XL et al (2006) J At Mol Phys: Suppl 48 (Relative deviation is less than 6.59×10^{-5})
83. Yin Z, Nie YC, Zhang GY et al (2006) J Tsinghua Univ: Nat Sci Ed 46:2037. (Relative deviation $\leq 1.71 \times 10^{-5}$)
84. Zhang GY, Xue LP, Cheng Y et al (2004) J At Mol Phys 21:411
85. Cheng Y, Zhang GY, Xue LP et al (2005) J Shangqiu Norm Univ 21:22
86. Xu GX, Li LM, Wang DM (1985) Basic principle and ab initio method of quantum chemistry, vol 2. Science Press, Beijing
87. Murrell JN, Kettle SFA, Tedder JM (1978) Valence theory (trans: Wen Z, Yao W et al). Science Press, Beijing
88. Zheng NW, Wang T et al (2001) Calculation of the transition probability for C (i–iv). J Opt Soc Am B 18:1395–1409
89. Zhang TY, Zheng NW, Ma DZ (2009) Theoretical calculations of transition probabilities and oscillator strengths for Ti III and Ti IV. Int J Quantum Chem 109(2):145–159
90. Zheng NW, Wang T et al (2000) Theoretical calculation of transition probability for N atom and ions. J Chem Phys 112:7042
91. Zheng NW, Wang T et al (2002) Theoretical study of transition probability for oxygen atom and ions. J Phys Soc JPN 71:1672–1675
92. Zheng NW, Wang T (2003) Spectrochim Acta B 58:27; (2003) 58:1319
93. Zheng NW, Wang T et al (1999) Study of transition probability of low states of alkali metal atoms with WBEPM Theory. J Phys Soc JPN 68:3859–3862

94. Zheng NW, Sun YJ, Wang T et al (2000) Transition probability of lithium atom and lithiumlike ions with weakest bound electron wave functions and coupled equations. Int J Quantum Chem 76:51
95. Zhang TY, Zheng NW (2009) Theoretical study of energy levels and transition probabilities of boron atom. Acta Phys Pol, A 116(2):141
96. Zheng NW, Wang T et al (2000) Transition probability of Cu I, Ag I, and Au I from weakest bound electron potential model theory. J Chem Phys 113:6169.s
97. Zheng NW, Wang T et al (2001) Transition probabilities for Be I, Be II, Mg I, and Mg II. At Data Nucl Data Tables 79:109–141
98. Zheng NW, Wang T (2002) Theoretical resonance transition probabilities and lifetimes for atomic nitrogen. Chem Phys 282:31–36
99. Zheng NW, Wang T (2002) Astrophys J Suppl Ser 282:31
100. Fan J (2003) Theoretical study on the energy levels and the transition properties for second-period and third-period atoms. University of Science and Technology of China, Hefei
101. Zheng NW, Fan J (2003) Theoretical study of energy levels and transition probabilities of singly ionized aluminum (Al II). J Phys Soc JPN 72:3091–3096
102. Fan J, Zheng NW (2004) Oscillator strengths and transition probabilities for Mg-like ions. Chem Phys Lett 400:273–278
103. Fan J, Zheng NW et al (2007) Calculations for spin-allowed transitions between energy levels above the 3s3p state in Si III. Chin J Chem Phys 20:265–272
104. Wang T (2003) Theoretical study on the transition, ionization and energy levels of many-electron atoms and ions within WBEPM Theory. University of Science and Technology of China, Hefei
105. Zeng J (1993) Introduction to quantum mechanics. Science Press, Beijing
106. Hoans-Binh D (1993) Multiplet oscillator strengths for excited atomic magnesium. Astron Astrophys Suppl Ser 97:769–775
107. Fischer CF (1975) Can J Phys 53:338; (1975) 53:184
108. Fischer CF, Tachiev G, Irimia A (2006) Relativistic energy levels, lifetimes, and transition probabilities for the sodium-like to argon-like sequences. At Data Nucl Data Tables 92:607–812
109. Fischer CF, Ralchenko Y (2008) Multiconfiguration Dirac–Hartree–Fock energies and transition probabilities for 2p4(P3), 3d–2p4(P3)4f transitions in Ne II. Int J Mass Spectrom 271:85–92
110. Fischer CF (2006) Some improved transition probabilities for neutral carbon. J Phys B: At Mol Opt Phys 39:2159
111. Moccia R, Spizzo P (1985) J Phys B 18:3537
112. Hibbert A, Biement E, Godefroid M et al (1993) Accurate F values of astrophysical interest for neutral carbon. Astron Astrophys Suppl Ser 99:179–204
113. Fawcett BC (1987) Oscillator strengths of allowed transitions for C I, N II, and O III. At Data Nucl Data Tables 37:411–418
114. Tong M, Fischer CF, Sturesson L (1994) J Phys B 27:4819
115. Nahar SN (1998) Phys Rev A 58:3766
116. Velasco M, Lavin C, Martin I (1997) J Quant Spectrosc Radiat Transfer 57:509
117. Migdalek J, Baylis WE (1978) J Phys B 11:L497
118. Safronova UI, Safronova AS, Beiersdorfer P (2008) Relativistic many-body calculations of lifetimes, rates, and line strengths of multipole transitions between $3l^{-1} 4l'$ states in Ni-like ions. Phys Rev A 77:032506
119. Johnson WR, Safronova UI (2007) Revised transition probabilities for Fe XXV: relativistic CI calculations. At Data Nucl Data Tables 93:139–147
120. Safronova UI, Cowan TE, Safronova MS (2006) Relativistic many-body calculations of energies, E2, and M1 transition rates of $4s^24p$ states in Ga-like ions. Phys Lett A 348:293–298
121. Murakami I, Safronova UI, Vasilyev AA et al (2005) Excitation energies, radiative and autoionization rates, dielectronic satellite lines, and dielectronic recombination rates to excited states for B-like oxygen. At Data Nucl Data Tables 90:1–74

122. Safronova UI, Johnson WR, Shlyaptseva A et al (2003) Dynamic transition in driven vortices across the peak effect in superconductors. Phys Rev A 67:052507
123. Johnson WR, Savukov IM, Safronova UI et al (2002) Astrophys J Suppl Ser 141:543
124. Borschevsky A, Eliav E, Ishikawa Y et al (2006) Atomic transition energies and the variation of the fine-structure constant α. Phys Rev A 74:062505
125. Dzuba VA, Ginges JSM (2006) Calculations of energy levels and lifetimes of low-lying states of barium and radium. Phys Rev A 73:032503
126. Correge G, Hibbett A (2004) Transitions in C II, N III, and O IV. At Data Nucl Data Tables 86:19–34
127. Safronova MS, Williams CJ, Clark CW (2004) Relativistic many-body calculations of electric-dipole matrix elements, lifetimes, and polarizabilities in rubidium. Phys Rev A 69:022509
128. Keenan FP, Harra LK, Aggarwal KM et al (1992) Astrophys J 385:375
129. Kulaga D, Migdalek J, Bar O (2001) Transition probabilities and lifetimes in neutral barium. J Phys B 34:4775
130. Seijo L, Barandiaran Z, Harguindey E (2001) The ab initio model potential method: lanthanide and actinide elements. J Chem Phys 114:118
131. Cohen S, Aymar M, Bolovinos A et al (2001) Experimental and theoretical analysis of the 5pnpJ = 0e, 1e, 2e autoionizing spectrum of Sr. Eur Phys J D 13:165–180
132. Seaton MJ (1998) Oscillator strengths in Ne I. J Phys B 31:5315
133. Rohrlich F (1959) Astrophys J 129:441; (1959) 129:449
134. Racah G (1942) Phys Rev 62:438; (1943) 63:367
135. Fuhr JR, Martin WC, Musgrove A et al (1996) NIST Atomic Spectroscopic Database, Version 2.0. http://physics.nist.gov (Select "Physical Reference Data")
136. Lindgard A, Nielsen SE (1975) Numerical approach to transit probabilities in the Coulomb approximation: Be II and Mg II Rydberg series. J Phys B 8:1183
137. Lindgard A, Nielsen SE (1977) Transition probabilities for the alkali isoelectronic sequences Li I, Na I, K I, Rb I, Cs I, Fr I. At Data Nucl Data Tables 19:533–633
138. Theodosiou CE (1984) Lifetimes of alkali-metal—atom Rydberg states. Phys Rev A 30:2881
139. Theodosiou CE (1980) Minima in the emission oscillator strengths of alkali Rydberg states. J Phys B 13:L1
140. Zheng NW, Sun YJ et al (1999) The radial expectation values for ground neutral atom ($Z = 2$–54). Acta Phys Chim Sin 15:443–448
141. Sun YJ, Wang T, Zheng NW (1998) Comput Appl Chem 5:369
142. Celik G et al (2007) Comparison of transition probabilities calculated using different parameters on WBEPM theory for some p-d and d-p transitions in excited atomic nitrogen. Int J Quantum Chem 107:495–500
143. Evans EH, Day JA, Fisher A et al (2004) Atomic spectrometry update. Advances in atomic emission, absorption and fluorescence spectrometry and related techniques. Anal At Spectrom 19:775–812
144. Celik G, Akin E, Kilic HS (2006) The theoretical calculation of transition probabilities for some excited p-d transitions in atomic nitrogen. Eur Phys J D 40:325–330
145. Celik G, Ates S (2007) The calculation of transition probabilities for atomic oxygen. Eur Phys J D 44:433–437
146. Bridges JM, Wiese WL (2007) Experimental study of weak intersystem lines and related strong persistent lines of Ne II. Phys Rev A 76:022513
147. Baclawski A, Wujec T, Musielok J (2006) Line strength measurements for near-infrared intersystem transitions of NI. Eur Phys J D 40:195–199
148. Hikosaka Y, Lablanquie P, Shigemasa E (2005) Efficient production of metastable fragments around the 1s ionization threshold in N2. J Phys B: At Mol Opt Phys 38:3597
149. Celik G (2007) The calculation of transition probabilities between individual lines for atomic lithium. J Quant Spectrosc Radiat Transf 103:578–587
150. Fvet V, Quinet P, Biemont E et al (2006) Transition probabilities and lifetimes in gold (Au I and Au II). J Phys B: At Mol Opt Phys 39:3587

151. Baclawski A, Musielok J (2007) Testing recent calculations of astrophysical relevant infrared NI line strengths by arc emission measurements. Eur Phys J Spec Top 144:221–225
152. Baclawski A, Wujec T, Musielok J (2007) Measurements of selected NI multiplet strength ratios and comparison with recent calculations. Eur Phys J D 44:427–431
153. Wang W, Cheng XL, Yang XD et al (2006) Calculation of wavelengths and oscillator strengths in high-Z Mg-like ions. J Phys B: At Mol Opt Phys 39:519
154. Santos JP, Madruga C, Parente F et al (2005) Relativistic transition probabilities for F-like ions with $10 \leqslant Z \leqslant 49$. Nucl Inst Methods Phys Res B 235:171–173
155. Mahmood S, Arnin N, Sami-ul-Haq et al (2006) Measurements of oscillator strengths of the $2p^5(^2P_{1/2})nd\ J = 2, 3$ autoionizing resonances in neon. J Phys B: At Mol Opt Phys 39:2299
156. Yang ZH (1994) J At Mol Phys 11:330
157. Yang ZH (1994) J At Mol Phys 11:445
158. Yang ZH, Su YW (1996) Acta Photonica Sin 25:783
159. Zheng NW, Wang T, Ma DX et al (2004) Weakest bound electron potential model theory. Int J Quantum Chem 98:281–290
160. (a) Ma DX, Zheng NW, Fan J (2005) Variational treatment on the energy of the He-sequence ground state with weakest bound electron potential model theory. Int J Quantum Chem 105:12–17; (b) Ma DX (2005) The development of WBEPM Theory and the study of its application. University of Science and Technology of China, Hefei; (c) Zheng NW, Zhang TY (2009) Calculation of He atom ground state using double generalized Laguerre polynomial in the WBEPM Theory. Acta Phys-Chim Sin 25:1093; (d) Zheng NW, Ma DX et al Perturbation treatment on total energies and ionization energies of He-like series in WBEPM Theory (unpublished)
161. King FWJ (1997) Progress on high precision calculations for the ground state of atomic lithium. Mol Struct (Theochem) 400:7–56
162. King FW (1999) High-precision calculations for the ground and excited states of the lithium atom. Adv At Mol Opt Phys 40:57–112
163. Drake GWF (1999) High precision theory of atomic helium. Phys Scr T 83:83
164. Drake GWF (2002) Ground-state energies for helium, H^-, and Ps^-. Phys Rev A 65:054501
165. Rahal H, Gombert MM (1997) Theoretical derivation of bound and continuum states of two-electron ions with model potential. J Phys B: At Mol Opt Phys 30:4695
166. Springborg M (2000) Methods of electronic-structure calculations. Wiley, Chichester, pp 100–101
167. Veszpremi T, Feher M (1999) Quantum chemistry: fundamentals to applications. Kluwer Academic/Plenum Publishing, Dordrecht, p 113
168. Kim YK (1967) Relativistic self-consistent-field theory for closed-shell atoms. Phys Rev 154:17
169. Levine IN (2000) Quantum chemistry, 5th edn. Prentice Hall, Opper Saddle River, NJ
170. Lide DR (2000) CRC handbook of chemistry and physics. CRC Press Inc., Boca Raton
171. Foresman JB, Frisch A (1996) Exploring chemistry with electronic structure methods, 2nd edn. Gaussian Inc., Pittsburgh
172. Pilar FL (1968) Elementary quantum chemistry. McGraw-Hill, Inc., New York, p 248
173. Wen GW, Wang LY, Wang RD (1991) Chin Sci Bull 36:547
174. Pauling L (1966) The nature of the chemical bond (trans: Lu JX, Huang YC, Zeng GZ et al). Shanghai Scientific and Technical Publishers, Shanghai, p 3
175. Pauling L (1966) The nature of the chemical bond (trans: Lu JX, Huang YC, Zeng GZ et al). Shanghai Scientific and Technical Publishers, Shanghai, p 79
176. Pauling L (1966) The nature of the chemical bond (trans: Lu JX, Huang YC, Zeng GZ et al). Shanghai Scientific and Technical Publishers, Shanghai, pp 79–86.
177. Mulliken RS (1934) J Chem Phys 2:782; (1935) 3:573. Mulliken pointed out that the average value of ionization energy of an atom and electron affinity should be the measurement of attraction between neutral atoms and electrons. Since the magnitude of ionization energy reflects the difficulty of losing electrons, while the magnitude of electron affinity reflects the difficulty of getting electrons. So Mulliken used this average value as electronegativity scale,

i.e. $X_M = (I + A)/2$, where X_M represents Mulliken electronegativity scale, I and A represents ionization energy of neutral atoms and electron affinity, respectively
178. Allred AL, Rochow ER (1958) J Inorg Nucl Chem 5:264. Allred-Rochow electronegativity scale starts from Coulomb force, i.e., where F represents the Coulomb attractive force between the nucleus and outer shell valence electrons, Z^* is the effective nuclear charge acting on valence electrons, r is the covalent radius of the atom. Z^* can be calculated by Slater rules. By fitting to Pauling electronegativity, the formula for electronegativity X is finally obtained, and
179. Gordy W (1946) Phys Rev 69:604. Gordy's contribution is to give the values of electronegativity at different valence states
180. Sanderson RT (1952) J Chem Educ 29:539; (1954) 31:2; 238
181. Phillips JC (1969) Covalent bonding in crystals, molecules and polymers. University of Chicago Press, Chicago
182. John JS, Bloch AN (1974) Quantum-defect electronegativity scale for nontransition elements. Phys Rev Lett 33:1095
183. Allen LC (1989) Electronegativity is the average one-electron energy of the valence-shell electrons in ground-state free atoms. J Am Chem Soc 111:9003–9014
184. Zhang YH (1982) Electronegativities of elements in valence states and their applications. 1. Electronegativities of elements in valence states. Inorg Chem 21:3886–3889. Y. H. Zhang's electronegativity scale of elements in valence states also starts from Coulomb forces, i.e. where Z^* is the effective nuclear charge, r is the covalent radius. Then use the expression for ionization energy I_1: to derive , and by substituting it into the previous equation, is obtained. Finally use Pauling electronegativity and to make a plot, and the following formula can be obtained by fitting: . In the process of treatment, the value of I_1 was taken from Refs. [6] and [7] of this paper and the effective principle quantum number n^* was taken from Ref. [7] of this paper, but in this paper the author of Ref. [7] was written as Zhen Nengwu by mistake and it should be Zheng Nengwu. This scale gives electronegativity of the same element at different valence states and can be further related to the strength of soft and hard acids.
185. Zheng NW, Li GS (1994) J Phys Chem 98:3964. N. W. Zheng and G. S. Li gave the nuclear potential scale to the electronegativity of the weakest bound electron
186. Liu ZX (1942) J Chin Chem Soc 9:119
187. Sun CE (1943) J Chin Chem Soc 10:77
188. Li SJ (1957) Acta Chim Sin 23:234
189. Gao XH (1961) Acta Chim Sin 27:190
190. Yuan HJ (1964) Acta Chim Sin 30:341; (1965) 31:536
191. Parr RG, Donnelly RA, Levy M et al (1978) J Chem Phys 69:4431
192. Parr RG, Yang WT (1989) Density-functional theory of atoms and molecules. Oxford University Press, Inc., New York. The given chemical potential is the negative value of Mulliken electronegativity, i.e. $\mu = -X_M$
193. Iczkowski R et al (1961) J Am Chem Soc 83:3547. The formula was given
194. Johnson KH (1977) Int J Quantum Chem 11:39. The orbital electronegativity was defined as $\mu_i = -\frac{\partial E}{\partial n_i} = -\epsilon_i$. This formula indicates that when the ith orbital has dn_i more electrons, the decreasing rate of its energy is μ_i. The larger μ_i, the stronger the capability of accepting electrons of this orbital. The negative value of the orbital energy in X_a method is equal to the orbital electronegativity μ_i.
195. Tang AY, Yang ZZ, Li QS (1982) Quantum chemistry. Science Press, Beijing
196. Pearson RG (1990) Absolute electronegativity and absolute hardness // Maksic ZB The concept of chemical bond. Springer-Verlag, Berlin, pp 45–76. R. G. Pearson gave the concept of absolute electronegativity and absolute hardness. Absolute electronegativity and absolute hardness
197. Huheey J (1965) J Phys Chem 69:3284
198. Yang P (2007) Structural parameters of molecules and the rule of their association with physical properties. Science Press, Beijing. In this book, a summary of the research on electronegativity was systematically elaborated

199. Yang P, Gao XH (1987) Property-structure-chemical bond. Higher Education Press, Beijing
200. Li GS, Zheng NW (1994) Acta Chim Sin 52:448
201. Gazquez JL, Ortiz E (1984) J Chem Phys 81:2741
202. Huheey JE (1983) Inorganic chemistry: principles of structure and reactivity, 3rd edn. Harper International Si Edition, Cambridge
203. Pearson RG (1990) Absolute electronegativity and absolute hardness // Maksic ZB The concept of the chemical bond. Springer-Verlag, Berlin
204. Abrland S, Davies NR (1958) Quant Rev 19:265
205. Dai AB et al (1987) Coordination chemistry. Science Press, Beijing
206. Pearson RG (1963) J Am Chem Soc 85:3533
207. Pearson RG Hard and soft acids and bases. Dowden, Hutchinson and Ross, Stroudsburg, Pa.
208. Inorganic chemistry group in the Chemistry Department at Nanjing University (1976) Chemistry 6:47
209. Pearson RG (1968) Hard and soft acids and bases, HSAB, part II: underlying theories. J Chem Educ 45:643
210. Zhang YH (1982) Electronegativities of elements in valence states and their applications. 1. Electronegativities of elements in valence states. Inorg Chem 21:3886–3889
211. Drago RS, Kabler RA (1972) Quantitative evaluation of the HSAB [hard-soft acid-base] concept. Inorg Chem 11:3144–3145; Drago RS (1973) Pearson's quantitative statement of HSAB [hard-soft acid-base]. Inorg Chem 12:2211–2212
212. Klopman G (1968) Chemical reactivity and the concept of charge- and frontier-controlled reactions. J Am Chem Soc 90:223–234
213. Dai AB (1978) Chemistry 1:26
214. Pedersen CJ (1967) Cyclic polyethers and their complexes with metal salts. J Am Chem Soc 89:2495–2496
215. Cram DJ, Cram JM (1974) Host-guest chemistry: complexes between organic compounds simulate the substrate selectivity of enzymes. Science 183:803–809; Cram DJ, Cram JM (1978) Design of complexes between synthetic hosts and organic guests. Acc Chem Rev 11:8–14
216. Lehn JM (1978) Cryptates: inclusion complexes of macropolycyclic receptor molecules. Pure Appl Chem 50:871; Lehn JM (1988) Supramolecular chemistry—scope and perspectives molecules, supermolecules, and molecular devices (Nobel Lecture). Angew Chem Int Ed Engl 27:89–112; Lehn JM (1993) Supramolecular chemistry. Science 260:1762–1763
217. Schmidt GMJ (1971) Photodimerization in the solid state. Pure Appl Chem 27:647–678
218. Wells AF (1984) Structural inorganic chemistry, 5th edn. Oxford University Press, Oxford
219. Robson R, Abrahams BF, Batten SR et al (1992) In supramolecular architecture // Bein T Acs symposium series 499. American Chemical Society, Washington, D.C.
220. Batten SR, Robson R (1998) Interpenetrating nets: ordered, periodic entanglement. Angew Chem Int Ed 37:1460–1494
221. Zaworotko MJ (1994) Crystal engineering of diamondoid networks. Chem Soc Rev 23:283–288
222. Moulton B, Zaworotko MJ (2001) From molecules to crystal engineering: supramolecular isomerism and polymorphism in network solids. Chem Rev 101:1629–1658
223. Yaghi OM, Sun Z, Richardson DA et al (1994) Directed transformation of molecules to solids: synthesis of a microporous sulfide from molecular germanium sulfide cages. J Am Chem Soc 116:807–808
224. Gardner GB, Venkataraman D, Moore JS et al (1995) Spontaneous assembly of a hinged coordination network. Nature 374:792–795
225. Fujita M, Kwon YJ, Washizu S et al (1994) Preparation, clathration ability, and catalysis of a two-dimensional square network material composed of cadmium(II) and 4,4'-bipyridine. J Am Chem Soc 116:1151–1152
226. Orr GW, Barbour LJ, Atwood JL (1999) Controlling molecular self-organization: formation of nanometer-scale spheres and tubules. Nature 285:1049–1052

References

227. Blake AJ, Champness NR, Chung SSM et al (1997) In situ ligand synthesis and construction of an unprecedented three-dimensional array with silver(i): a new approach to inorganic crystal engineering. Chem Commun 1675–1676
228. Pan L, Huang XY, Li J et al (2000) Novel single- and double-layer and three-dimensional structures of rare-earth metal coordination polymers: the effect of lanthanide contraction and acidity control in crystal structure formation. Angew Chem Int Ed 39:527–530
229. Pan L, Xheng NW, Wu YG et al (2001) Synthesis, characterization and structural transformation of a condensed rare earth metal coordination polymer. Inorg Chem 40:828–830
230. Sun JY, Wong LH, Zhou YM et al (2002) QMOF-1 and QMOF-2: three-dimensional metal-organic open frameworks with a quartzlike topology. Angew Chem Int Ed 41:4471–4473
231. Sun YQ, Zhang J, Chen YM et al (2005) Porous lanthanide-organic open frameworks with helical tubes constructed from interweaving triple-helical and double-helical chains. Angew Chem Int Ed 44:5814–5818
232. Muller-Buschbaum K, Mokaddem Y, Schappacher FM et al (2007) $\{Eu^{II}N_{12}\}$: a homoleptic framework containing EuIIN12 icosahedra. Angew Chem Int Ed 46:4385–4387
233. Rao CNR, Nafarajan S, Vaidhyanathan R (2004) Metal carboxylates with open architectures. Angew Chem Int Ed 43:1466–1496
234. Xu HT, Zheng NW, Jin XL et al (2002) Rare earth ion center in distorted square anti-prism coordination environment: a novel three-dimensional coordination polymer with channels. Chem Lett 31:350–351
235. Xu HT, Zheng NW, Jin XL et al (2002) Reactions of zirconacyclopentadienes with CO, CN, and NN moieties with electron-withdrawing groups: formation of six-membered heterocycles. Chem Lett 124:1144–1145
236. Wu YG, Zheng NW, Yang RY et al (2002) From condensed coordination structure to open-framework by modifying acid ligand. J Mol Struct 610:181–186
237. Xu HT, Zheng NW, Jin XL et al (2003) A new microporous structure constructed by a lanthanide-carboxylate coordination polymer. J Mol Struct 654:183–186
238. Xu HT, Zheng NW, Jin XL et al (2003) Assembly of lanthanide coordination polymers with one dimensional channels. J Mol Struct 655:339–342
239. Pan L, Zheng NW, Wu YG et al (1999) Synthesis of one-and two-dimensional zinc and cadmium complexes with 4, 4'-bipy. J Coord Chem 47:269–277
240. Pan L, Zheng NW, Wu YG (1999) Chin J Struct Chem 18:41
241. Pan L, Zheng NW, Zhou XY et al (1998) Catena-Poly[[tetrakis(imidazole-N^3)copper(II)]-μ-(dichromato-O:O')]. Acta Cryst C54:1802–1804
242. Pan L, Zheng NW, Wu YG et al (1999) Poly[mercury(II)-μ-4,4'-bipyridine-di-μ-bromo] Acta Cryst C55:343–345
243. Pan L, Zheng NW, Wu YG et al (1999) J Coord Chem 47:269
244. Pan L, Zheng NW, Wu YG et al (1999) J Coord Chem 47:551
245. Pan L, Zheng NW, Wu YG et al (2000) A three-dimensional structure, four-connected network of the zeolite Li–A(BW) topological frame sustained by [Hg(hmt)1/2Br 2] units via secondary bonding. Inorg Chim Acta 303:121–123
246. Xu HT, Zheng NW, Xu HH et al (2001) The study on single crystal structure of [Zn(Hpdc)2(H_2O)2]·$2H_2O$ (Hpdc−=2,5-pyridinedicarboxylic acid group). J Mol Struct 597:1–5
247. Xu HT, Zheng NW, Xu HH et al (2002) Synthesis and studies on single crystal structure of [Ni(3,5-pdc)·H_2O] (3,5-pdc=3,5-pyridinedicarboxylic acid). J Mol Struct 606:117–122
248. Xu HT, Zheng NW, Jin XL et al (2003) Channel structure of diaquasesqui (2-aminoterephthalato) dysprosium(III) dihydrate. J Mol Struct 646:197–199
249. Xu HT, Zheng NW, Yang RY et al (2003) The effect of the ligand's symmetry on assembly of the coordination polymer [Mn($C_6N_2H_2O_4$)]: a new coordination polymer with channel structures. Inorg Chim Acta 349:265–268
250. Chen GJ, Ouyang Y, Yan SP et al (2008) Magnetic property of a 1D triazacycloalkanes Ni(II) with chiral helical structure complex $\{[Ni[12]aneN3](\mu_{N,S}\text{-SCN})(SCN)\}n$. Inorg Chem Commun 11:138–141

251. Xu JY, Tian JL, Zhang QW et al (2008) Tetranuclear pyrophosphate-bridged Cu(II) complex with 2,2'-dipyridylamine: crystal structure, spectroscopy and magnetism. Inorg Chem Commun 11:69–72
252. Luo J, Zhou XG, Gao S et al (2004) Synthesis and characterization of copper(II) azide complexes containing polyamines as co-ligands. Polyhedron 23:1243–1248
253. Yang RY, Zheng NW, Xu HT et al (2002) Chin Sci Bull 47:1546
254. Yang RY, Zheng NW, Xu HT et al (2003) Chemistry 7:492
255. Pan L (2000) Design Synthesis and characterization of coordination polymers. University of Science and Technology of China, Hefei
256. Xu GX, Gao S, Huang CH et al (1993) Prog Nat Sci 3:1
257. Huang CH (1997) Rare earth coordination chemistry. Science Press, Beijing, p 1997
258. Long DL, Blake AJ, Champness NR et al (2001) Lanthanum coordination networks based on unusual five-connected topologies. J Am Chem Soc 123:3401–3402
259. Wu YG (2001) The synthesis, structure and properties of coordination polymers. University of Science and Technology of China, Hefei
260. Xu HT (2002) Synthesis, structural characterization and study of properties of rare earth or transition metal diacid coordination polymers. University of Science and Technology of China, Hefei
261. Xu HT, Zheng NW, Xu HH et al (2002) Mn(II) ion centers in novel coordination environment: a new two-dimensional coordination polymer [Mn(3,5-pdc)·2H$_2$O]. J Mol Struct 610:47–52

Representation Publications

Monographs

(1) Zheng N W (2010) The Weakest Bound Electron Theory and Its application [M]. Hefei: University of Science and Technology of China Press
(2) Zheng N W (1988) A New Outline to Atomic Theory [M]. Nanjing: Nanjing Education Press

Papers

(1) Zheng N W (1977) Chinese Science Bulletin 22: 531.
(2) Zheng N W (1985) Chinese Science Bulletin 30: 1801; Zheng N W (1986) KEXUE TONGBAO 31: 1238.
(3) Zheng N W (1986) Chinese Science Bulletin 31: 1316; Zheng N W (1987) KEXUE TONGBAO 32: 1263.
(4) Zheng N W (1987) Chinese Science Bulletin 32: 354; Zheng N W (1988) KEXUE TONGBAO 33: 916.
(5) Zheng N W, Xin H W (1991) Successive ionization potentials of 4fn electrons within 'WBEPM' theory'. J. Phys. B: At. Mol. Opt. Phys. 24: 1187.
(6) Zheng N W, Li G S (1994) Electronegativity: Average Nuclear Potential of the Valence Electron. J. Phys. Chem. 98: 3964–3966
(7) Zheng N W, Wang T et al. (2000) Transition probability of Cu I, Ag I, and Au I from weakest bound electron potential model theory. J. Chem. Phys. 113: 6169.
(8) Zheng N W, Ma D X et al. (2000) An efficient calculation of the energy levels of the carbon group. J. Chem. Phys. 113: 1681.
(9) Zheng N W, Wang T et al. (2000) Theoretical calculation of transition probability for N atom and ions. J. Chem. Phys. 112: 7042.
(10) Zheng N W, Sun Y J (2000) Science in China, Ser. B-Chem. 30: 15; Zheng N W, Sun Y J. Sun Y J (2000) Science in China, Ser. B-Chem. 43: 113.
(11) Pan L, Huang X Y, Li J, Wu Y G, Zheng N W (2000) Novel Single- and Double-Layer and Three-Dimensional Structures of Rare-Earth Metal Coordination

Polymers: The Effect of Lanthanide Contraction and Acidity Control in Crystal Structure Formation. Angew. Chem. Int. Ed. 39: 527–530
(12) Zheng N W, Zhou T et al. (2000) Analysis of the bound odd-parity spectrum of krypton by weakest bound electron potential model theory. Chem. Phys., 2000, 258: 37–46
(13) Zheng N W, Wang T, et al. (2001) TRANSITION PROBABILITIES FOR Be I, Be II, Mg I, AND Mg II. At. Data Nucl. Data Tables 79: 109–141
(14) Zheng N W, Wang T, et al. (2001) Calculation of the transition probability for C $_{(I-IV)}$. J. Opt. Soc. Am. B-Opt. Phys. 18: 1395–1409
(15) Pan L, Zheng N W, et al. (2001) Synthesis, Characterization and Structural Transformation of A Condensed Rare Earth Metal Coordination Polymer. Inorg. Chem. 40: 828–830
(16) Zheng N W, Zhou T et al. (2002) Ground-state atomic ionization energies for $Z = 2 - 18$ and up to 18 electrons. Phys. Rev. A 65: 052,510.
(17) Zheng N W, Wang T (2002) Radiative Lifetimes and Atomic Transition Probabilities for Atomic Carbon and Oxygen. Astrophys. J. Suppl. Ser. 143: 231.
(18) Wu Y G, Zheng N W et al. (2002) From condensed coordination structure to open-framework by modifying acid ligand. J. Mol. Struct. 610: 181–186
(19) Zheng N W, Wang T (2003) Systematical study on the ionization potential of excited states in carbon-like sequence. Chem. Phys. Letts. 376: 557–565
(20) Zheng N W, Wang T et al. (2004) Weakest bound electron potential model theory. Int. J. Quantum Chem. 98: 281–290
(21) Ma D X, Zheng N W et al. (2004) Theoretical Analysis on 3dnl J = 1e–5e Autoionizing Levels in Ca. J. Phys. Chem. Ref. Data, 2004, 33: 1013
(22) Fan J, Zheng N W (2004) Oscillator strengths and transition probabilities for Mg-like ions. Chem. Phys. Letts. 400: 273–278
(23) Ma D X, Zheng N W et al. (2005) Variational treatment on the energy of the He-sequence ground state with weakest bound electron potential model theory. Int. Quantum Chem. 105: 12–17
(24) Zhang T Y, Zheng N W, Ma D X (2007) Theoretical calculation of energy levels of Sr I. Phys. Scr. 75: 763.
(25) Zhang T Y, Zheng N W, Ma D X (2009) Theoretical calculations of transition probabilities and oscillator strengths for Ti III and Ti IV. Int. J. Quantum Chem. 109: 145–159

Textbooks and Teaching Reference Books

(1) Zheng N W, Liu Q L, Liu S H (1988) Principles of Inorganic Chemistry [M]. Hefei: University of Science and Technology of China Press
(2) Zheng N W, Zhang H L, Zhao W R (1985) The Physical Concept of Chemical Bond [M]. Hefei: Anhui Science and Technology Press

Popular Science Books

(1) Zheng N W (1981) The First Element [M]. Hefei: Anhui Science and Technology Press. This book was translated by Ma L L into Korean and published.
(2) Zheng N W (1983) The family of air [M]. Beijing: China Children Press. This book was published again as: Zheng N W, Gu S X (1991) Know Air and Water [M]. Taipei: Qian Qian Press. In 2000, *the book Air and Water* written by Zheng N W and Gu S X was included in the book *Chemistry Laboratory* by the editorial board of 21-century encyclopedia series for teenagers.

Postscript

I have been committed to constructing a new quantum theory based on wave-particle duality for many years. The weakest bound electron theory is a trial of this idea. Just as Mr. Xun Lu said, path is shown up only when thousands of people walk through. Now that we have left some footprints for a path in a field with no road, I hope it will become a road when more people walk in the future.

Index

A
A closed cycle, 26, 35
Additivity, 33
Analytical potential, 39, 44, 47, 55, 58, 101
Angular equation, 40
Atomic core, 26, 28–31, 37, 38, 104, 108, 114, 131, 188
Atomic energy level, 2, 8, 69, 97–102, 105–108, 113, 123
Aufbau process, 13, 24, 25
Autoionizing level, 106, 123

B
Borderline acid, 186
Borderline base, 186, 187

C
Coordination polymer, 96, 179, 185, 188–193, 195, 196
Crystal engineering, 189

D
Double generalized Laguerre function, 166, 169, 174, 175
Doubly excited state, 123

E
Electron affinity, 181, 185, 188
Electron configuration, 13, 14, 25, 78, 98, 108, 115, 123, 130, 140
Electronegativity, 64, 179–182, 185, 187, 188
Electron-electron repulsion energy, 5

Electron-nuclei attraction energy, 5
Energy eigenfunction, The, 4
Energy eigenvalue, 4, 14, 58
Energy eigenvalue equation, The, 4, 5

G
Γ(gamma) function, 48
Generalized Laguerre function, 42, 48, 65, 158, 175
Generalized Laguerre polynomial, 42, 44, 48, 51–53, 56, 57, 165, 166

H
Hard acid, 186, 188, 191
Hard base, 186, 187, 191
Hellmann-Feynman theorem, 54, 56
HSAB matching, 190
HSAB theory, 186

I
Identical particle
    indistinguishability of identical particles, 10, 23, 35
Incomplete gamma function, 51, 164
Integral
    integral of electron repulsive energy, 161, 163, 164
    integral of kinetic energy, 161, 162, 164
    integral of nuclear attractive energy, 161, 163, 164
Ionization energy
    differential law of ionization energy, the, 76, 79

first difference of ionization energy, the, 79, 85
second difference of ionization energy, the, 79
Ionization process, 22, 23
Isoelectronic sequence, 76–78
Iso-spectrum-level series, 78–86, 89, 90, 92, 93, 96, 102, 107, 158

J
Jj coupling, 8, 100, 108

L
Lagrange multiplier, 17, 18
LS coupling, 8, 100, 108, 115

M
Matrix element of radial operator r^k, 56
Mean value of radial operator r^k, 56
Multiplicity of the term, 8

N
NCA method, 133
NIST data, 132
Nuclear potential scale to electronegativity, 182

O
Operator
  kinetic energy operator, the, 5, 29
  one-electron Hamiltonian operator, 55
  potential energy operator, the, 5
Oscillator strength, 129–131, 133, 134, 142, 151, 156

P
Pauli exclusion principle, 10, 13, 25, 166
Penetrate, 39
Perturbed level series, 106, 114
Principle of the indistinguishability of identical particles, The, 10

Q
Quantum number
  Azimuthal quantum number, 7, 13, 42, 43, 170, 180
  magnetic quantum number, 7, 13
  principle quantum number, 7, 13, 42, 43, 63, 79, 84, 99, 100, 104, 105, 180

R
Radial equation, 40, 42
Radiative lifetime, 129–131, 133, 134, 143
Relaxation effect, 24, 25, 35, 167

S
Scattering state, 58, 59
Schrodinger equation, The
  one-electron Schrodinger equation, 11, 33, 34, 38, 45, 47, 55, 58, 62, 64, 162
  stationary-state Schrodinger equation, the, 3
  time-dependent Schrodinger equation, 3–5
Screening, 84, 167, 168
Separability, 21, 23, 35, 157
Single generalized Laguerre function, 160, 166, 169, 175, 179
Slater determinant, 11–13
Soft acid, 186, 191
Soft base, 186, 187, 191
Spectral branch, 8
Spectral term, 77, 78, 98, 99, 130, 156
Spectrum-level-like series, 100, 108–110, 112, 114, 115
Spherical harmonic, 40, 63, 163
Spin
  electron spin, 6, 8, 10, 11, 13, 160
  spin angular momentum, 6–8
  spin angular momentum along the magnetic field, 7
  spin function, 7, 13
  spin orbital, 6, 7, 11–13, 158, 166
  spin quantum number, 7, 9, 13
Spin-orbit coupling coefficient, 61, 62
Stationary Schrodinger equation, The, 4
Stationary state, 3
Stationary-state wave function, The, 3, 4
Subsystem, 23–26, 28, 31, 33
Successive ionization, 22–26, 28, 30, 35, 63, 65, 69, 70, 86, 91, 97, 158, 159, 185
Successive ionization energies of the 4fn electrons for the lanthanides, The, 91
Sum of one-electron Hamiltonians, The, 11, 64, 79

T

Total electron energy of the system, The, 161, 167
Transition probability
 probability of transition between electron configurations, the, 131
 probability of transition between energy levels, the, 131
 probability of transition between spectral terms, the, 131
Trial function, 160, 166, 169, 174

U

Uncertainty principle, The, 1, 2, 9
Unperturbed level series, 106

V

Variation method
 linear variation method, the, 14, 16
 Ritz method, 14, 15
Variation principle, The, 15

W

Wave function
 anti-symmetric wave function, 10
 symmetric wave function, 10
Weakest Bound Electron Potential Model Theory (WBEPM Theory), 37, 47, 48, 56–62, 65, 75, 83, 86, 88, 89, 91, 92, 95, 96, 98, 101, 105–107, 113, 114, 129, 130, 132, 133, 135, 157, 158, 160, 165–167, 169, 174–176
Weakest bound electron theory, 21, 37, 155
Weakest Bound Electron (WBE)
 Non-Weakest Bound Electron (N-WBE), 22, 23, 31, 45, 113
 one-electron energy of the weakest bound electron, the, 101
 one-electron Hamiltonian of the weakest bound electron, the, 34

Z

Zero state of total electronic energy, 35

CPSIA information can be obtained
at www.ICGtesting.com
Printed in the USA
LVHW051537120323
741461LV00005B/433